Green Mobility and the Environment: A Dialogue among Researchers

Green Mobility and the Environment: A Dialogue among Researchers

FELICE E. CORCIONE

Warrendale, Pennsylvania, USA

400 Commonwealth Drive
Warrendale, PA 15096-0001 USA
E-mail: CustomerService@sae.org
Phone: 877-606-7323 (inside USA and Canada)
724-776-4970 (outside USA)
FAX: 724-776-0790

Copyright © 2024 SAE International. All rights reserved.

No part of this publication may be reproduced, stored in a retrieval system, transmitted, in any form or by any means, electronic, mechanical, photocopying, recording, or otherwise, or used for text and data mining, AI training, or similar technologies, without the prior written permission of SAE. For permission and licensing requests, contact SAE Permissions, 400 Commonwealth Drive, Warrendale, PA 15096-0001 USA; e-mail: copyright@sae.org; phone: 724-772-4028.

Library of Congress Catalog Number 2024935664
http://dx.doi.org/10.4271/9781468608021

Information contained in this work has been obtained by SAE International from sources believed to be reliable. However, neither SAE International nor its authors guarantee the accuracy or completeness of any information published herein and neither SAE International nor its authors shall be responsible for any errors, omissions, or damages arising out of use of this information. This work is published with the understanding that SAE International and its authors are supplying information but are not attempting to render engineering or other professional services. If such services are required, the assistance of an appropriate professional should be sought.

ISBN-Print   978-1-4686-0801-4
ISBN-PDF   978-1-4686-0802-1
ISBN-epub   978-1-4686-0803-8

To purchase bulk quantities, please contact: SAE Customer Service

E-mail: CustomerService@sae.org
Phone: 877-606-7323 (inside USA and Canada)
        724-776-4970 (outside USA)
Fax: 724-776-0790

Visit the SAE International Bookstore at books.sae.org

Publisher
Sherry Dickinson Nigam

Product Manager
Amanda Zeidan

Production and Manufacturing Associate
Michelle Silberman

*To my grandchildren: Enrica, Anna, Andrea, and Luca,
protagonists and guardians of a green future.
To my wife, who shared her life with me in joy and in pain.*

Contents

Review I	x
Review II	xi
Review III	xii
Review IV	xiii
Preface	xiv
Acknowledgments	xviii
Introduction	xx

CHAPTER 1

Environmental Degradation — 1

1.1. Introduction	1
1.2. Causes of Environmental Degradation	5
1.3. Treatment of Municipal Solid Waste	6
1.4. Landfills	7
1.5. Waste-to-Energy Plants	8
1.6. Agriculture	9
1.7. Transport	9
1.8. Cost Analysis Between Diesel and Electric Cars	10

CHAPTER 2
European Union Proposals — 15

2.1. European Union Proposals: Fit for 55 — 15
2.2. New Proposals and Actions from Europe — 16
2.3. Euro 7 Regulation — 18

CHAPTER 3
The Rebirth of the Planet in the Name of Renewable Energy — 23

3.1. Introduction — 23
3.2. Solutions and Remedies — 24
3.3. Biofuels — 26
3.4. Synthetic Fuels and "E-Fuels" — 27
3.5. Circular Economy — 28

CHAPTER 4
Hydrogen as an Energy Carrier of the Future — 33

4.1. Hydrogen — 33
4.2. Different Types of Hydrogen — 35
4.3. Hydrogen from Ammonia — 36
4.4. Ammonia Engine — 37
4.5. The Announcement of Toyota — 37
4.6. Is it the Ammonia Era? — 38
4.7. Energy Revolution — 39
4.8. The Cost of Solar — 39
4.9. Progress of Nuclear Fusion — 40
4.10. New Generation: Waste to Energy Plants — 40

4.11. Anaerobic Digestion of the Wet Fraction of Waste	41
4.12. Storage and Transport of Hydrogen	42
4.13. Electrification of Mobility	43
4.14. Hydrogen Technology: Current Trends and Controversies	43
4.15. Low Emission Cars	48

CHAPTER 5
Mega-Trends in Technology 51

Epilogue	55
Conferences and Articles by the Author and Colleagues	57
Appendix A: Catania 2003: Electric-Fuel Cell Vehicles for Sustainable Mobility	59
Appendix B: Bressanone 2005: Exhibition and Test of Electric Vehicles	61
Appendix C: International Journal of Environmental Research and Public Health	65
Appendix D: Maurisi Vineyard Biodynamic Method	67
Appendix E: The Internal Combustion Engine	69
Index	75
About the Author	77

Review I

Once again Felice Corcione amazes us. This new work of his, which can be read effortlessly and in one go, has the enormous merit of "telling" about problems of a global dimension in simple ways and words. He tackles them with calmness, detachment, and acumen, which only the ones who are an absolute and total master of the subject can afford.

A meeting among friends, in front of a bottle and with a glass in hand, thus becomes the opportunity and the inspiration to reflect on serious and terribly looming problems in our contemporary society.

Felice masterfully summarizes the "dangerousness" and indicates, with rough lines, possible "implementations" to mitigate impending catastrophic effects.

Ferdinando Iannuzzi
Architect, hoplologist environmentalist, land scholar

Review II

The book *Green Mobility and the Environment: A Dialogue among Researchers* analyses the impact of pollution on the environment, mainly from means of transport, and the possible solutions to be adopted.

Despite being a very technical topic, there is an ease of reading that makes even the most difficult aspects understandable, so that even the ones who are not in the sector can understand what has been explained.

The language is professional, while still managing to be fluent and conversational, almost as if you were at a university lecture.

Nicoletta Bosio
Writer and proofreader

Review III

I congratulate you for the lucid and explanatory analysis, understandable even from a non-expert perspective, and I thank you for your quote and your gratitude. I do not fully agree with your pessimism about the timing of the new fission nuclear power plant. Studies and projects are underway not only in the US, but also in Turin, Italy by Stefano Buono's group, and developments will probably be seen in about 10 years. Ammonia is a molecule made up of three hydrogen and one nitrogen atoms. It does not contain carbon, and for this reason its use does not directly generate carbon dioxide emissions. While the high hydrogen content makes it possible to use it as a method of storing and transporting this element more easily than its pure form (which requires pressurized tanks at 700 bar), the first possibility, therefore, is to exploit ammonia for its greater ease of transport and storage, then to break it down into the elements composing it and to use hydrogen as fuel.

Roberto Gentili
Professor of the University of Pisa

Review IV

In a world where the very air we breathe is tainted by the consequences of unchecked industrialization and modern living, the imperative to confront air pollution looms larger than ever. This book seeks to unravel the intricacies of this environmental challenge and proposes a compelling solution: decarbonization. At its core, decarbonization is a transformative process designed to mitigate the adverse effects of air pollution by significantly reducing carbon emissions across industries.

As professor Corcione teaches us, hydrogen emerges as a key ally in the battle against pollution. Produced using renewable energy sources, it offers a clean alternative to conventional hydrogen production methods. This book explores the applications of hydrogen in transportation, with a specific focus on fuel cell vehicles that stand as emissaries of a future where mobility and environmental sustainability harmoniously coexist.

Technological advancements act as the guiding force, steering us toward a future where innovation meets environmental responsibility. The battle against air pollution demands not just awareness but actionable solutions grounded in a profound understanding of its myriad forms. The commitment to alternative, sustainable practices is our compass, and by embracing decarbonization and transitioning to zero-emission mobility, we collectively pave the way for a brighter, cleaner future—one where the air we breathe is liberated from the shackles of pollution.

La Forgia dei Libri
Editorial Services Agency

Preface

This book, written by Professor Felice E. Corcione, is intended to serve as a true operational guide regarding the interplay between "green mobility" and the environment, presenting viable pathways for a rapid transition from fossil fuels to hydrogen. It begins with the analysis of pollutants produced by endothermic engines (e.g., internal combustion engines) or waste-to-energy power stations. From there, the discussion shifts to the molecular basis of the greenhouse effect and an examination of recent global events concerning environmental disasters linked to climate change (e.g., prolonged droughts in Africa, freezing North American winters, sudden floods in our cities). These conversations can serve as rationale regarding the development of a "road map" as mobility shifts toward a "zero-emission" model enabled by "green" hydrogen. Drawing from ideas concerning the constitutional protection of the environment and human health, as well as strategies introduced by the European Union and international agencies, this work is designed to outline a proposal for the rebirth of the planet in the name of renewable energy and a circular economy model affecting every economic sector—from agriculture and manufacturing to transportation.

The heart of the text addressed hydrogen as an energy vector of the future, which one may ideally want as an opportunity for the present. Here, we discover how the fuel of the sun and other stars (a nod to the final verses of Dante's Divine Comedy) can be produced with "clean" systems, thanks to water electrolysis with electricity derived from renewable sources. National Aeronautics and Space Administration's (NASA's) Ames Research Center is an emblematic example, where two and a half tons of hydrogen were produced using just four megawatts of electricity. Without fear of contradicting Elon Musk and going against the current international trend of unconditional support for electric vehicles (EVs), Corcione has no doubts: hydrogen is the key to solve the energy challenge posed to humanity.

With an energy content higher than that of gasoline and natural gas, hydrogen can be used in fuel cells for land, air, and marine transport because of innovative solutions now available for its storage and transport. But there is more; the real surprise that the Director Emeritus of the National Research Council Motor Institute wants to lead us to is the possibility of using hydrogen in current internal combustion engines without complex modifications (beyond those already used for fuel natural gas and the addition of an ammonia photocatalyst developed by Rice University). An interesting and concrete example given by the author is the one of the University of New South Wales in Australia, where researchers have successfully transformed a traditional diesel engine into a dual fuel injection system using 90% hydrogen fuel. Thanks to Corcione, we are discovering that, in addition to its use in endothermic engines, hydrogen could soon power fuel cell vehicles. The path indicated by Corcione has solid scientific foundations, and we hope that it will soon become reality for the good of generations to come.

Prisco Piscitelli
Epidemiologist, Vice-president of the Italian
Society of Environmental Medicine

A life without research is not worth living.
—*Socrates*

Research understood as an instrument of knowledge and not as an object of competition and an instrument of power.
—*Rita Levi-Montalcini*

Acknowledgments

First, I would like to thank my friend Murli Iyer for encouraging me and helping me to publish this book. I thank my friends, with whom I share morning coffee at the Primavera bar and discussions during lively spring meetings in the farmhouse of my vineyard in Prata di Principato Ultra (AV). Thank you especially for encouraging me to write this book after my unfortunate health issues in 2019.

I thank my friends from Ischia and Panza for helping me to start swimming again and to write a book about the story of my family and professional life titled, *Il Ragazzo con il Cerchio*, (*The Boy with the Hoop*).

I thank Dr. Lena Auriemma for helping and supporting me during the writing of this text.

I thank the friends I met at the sulphurous swimming pool in Santa Cesarea Terme for sharing the hypotheses of transitioning toward a green environment and green mobility.

Heartfelt thanks go to my wife for the precious support she gave me in the intensive rereading and proofreading.

Precious thanks to my talented physiotherapist Vincenzo Penna who, with his professionalism, allowed me to resume the physical functions of the left side of my body after the tragic ischemic event of March 2019, allowing me to use both hands to write this text.

I warmly thank Professor Roberto Gentili of the University of Pisa for the useful suggestions on the use of ammonia in internal combustion engines.

Thanks to my violin teacher, Paolo Sasso, for encouraging me to move the fingers of my left hand to play the D major chord on my violin.

I thank my dear friend Prisco Piscitelli, careful researcher of the Euro Mediterranean Scientific and Biomedical Institute of Mesagne, for the precious discussions on environmental protection and sustainable mobility from the many meetings held in the 1980s at the Diocese of Acerra with the bishop Bishop Salvatore Giovanni Rinaldi.

Special thanks to my friends and colleagues, Peppino Police and Mariano Migliaccio, for their valid contribution on environmental and vehicular issues for the conversion to green transportation.

I thank Don Giovanni Rinaldi, bishop emeritus of Acerra, with whom I shared the concerns relating to the negative impact of the Acerra incinerator on the already tormented Marigliano-Acerrano territory.

A special thanks to my children, Carola and Giuseppe, who with their research and development activities have always shared with me the actions aimed at environmental protection.

Introduction

A Meeting between Researchers

First the earth then the car, my passion is green.[1]

Since 1974, I have always been involved in the fluid dynamics and combustion of endothermic engines, my aim being the minimization of harmful exhaust emissions by way of hydrogen and green hydrogen fuel cells as an alternative to fossil fuels.[2–8]

After carefully analyzing the current environmental situation and the repeated (albeit varying) proposals of the European Union from 2020 to 2050 regarding sustainable mobility,[1–10] I invited colleagues from the National Research Council (CNR) Motor Institute, experts from the CNR Institute of Science and Technology

---

[1] Interview with the newspaper "Il Mattino" di Napoli on October 12, 2003.
[2] Felice E. Corcione, "International Conference on the Environment: Waste, Water, Energy Living Bio," April 9–12, 2003, Catania.
[3] Felice E. Corcione, "Industrial Combustion and Polluting Emissions." Report presented at the Taranto conference, November 2018.
[4] Felice E. Corcione, "Decarbonization of the Puglia Region," Conference organized by President Michele Emiliano Rome November 2016.
[5] Industrial combustion and polluting emissions.
[6] Conference organized by Professor Alessandro Distante and Dr. Prisco Piscitelli at the Euro Mediterranean Biomedical Scientific Institute (ISBEM) in Mesagne (BR), July 2018.
[7] Characterization and control of a fuel cell propulsion system.
[8] Felice E. Corcione, "Let's Set Hydrogen in Motion." New Energy Sources, New Technologies, New Use and Diffusion Programs to Reduce Pollution, Conference at the Unione club in Marigliano, April 11, 2003.
[9] P. Corbo, Felice E. Corcione and F. Migliardini. Conference organized by the director of the CNR Motors Institute, 2004.
[10] EEA European Environment Agency. Trends and projections: Limited Increase in EU Emissions in a Context of Recovery from the Pandemic and Energy Crisis.

for Sustainable Energy and Mobility[11] from the University of Naples, and members of the corporate world to evaluate and compare the positive climate impacts of internal combustion engines vehicles equipped with new "green" technologies. Mariano Migliaccio, professor of internal combustion engines at the University of Naples (retired) welcomed the invitation with enthusiasm along with Giuseppe Police, former director of Istituto Motori (retired).

The meeting was held on Saturday, April 16, 2022, in Avellino, in Prata di Principato Ultra, at Via Vicinale Maurisi, home of my agricultural company—Vigna Maurisi—founded in 2011 to produce Greco di Tufo DOCG and Irpinia Aglianico POD grapes. They are grown in the biodynamic method, with utmost respect to environmental balance.

The chosen place is not random. The vineyard is near Pianodardine of Pratola Serra, home of Stellantis Hordain, recently formed by the merger between Fiat Chrysler Automobiles (FCA) and the French PSA Group. Stellantis Hordain is focusing its production on eco-friendly electric and hydrogen vehicles.

Upon reaching the farm, the full rebirth of flora and fauna is felt. At this time of year, the vines are in full bloom, the air is warm, the sky is clear of clouds, and the lizards run quickly to conquer a place in the sun. You can even see traces of the herds of wild boar looking for roots to feed on. Everything around favors scholarly dialogue among old research friends. Under the porch of the farmhouse, we enjoy glasses of Greco di Tufo and a good local cheese, during which I begin our noble conversation: our reason for being here. Starting with examination of the European perspective, I focus the attention on the following topics:

- Environmental degradation
- Treatment of municipal solid waste
- Proposals from the European Union
- Biofuels
- Synthetic fuels and e-fuels
- Hydrogen

---

[11] CNR- STEMS- Institute of Sciences and Technologies for Sustainable Energy and Mobility.

The objective of this book, the search for sustainable scientific citizenship, is the result of our dialogue. The aim is to arouse the curiosity of "citizen readers" regarding current and broad environmental problems and the theoretical and practical aspects of "green mobility." This is a contribution toward a future world of zero-emission mobility.

Let's not pretend that things will change if we keep doing the same things. A crisis can be a real blessing to any person, to any nation. For all crises bring progress. Creativity is born from anguish, just like the day is born form the dark night. It's in crisis that inventiveness is born, as well as discoveries made and big strategies. He who overcomes crisis, overcomes himself, without getting overcome.

—*Albert Einstein*

1

Environmental Degradation

1.1. Introduction

The environment is "the complex of conditions linked to time, place, material, social and cultural circumstances in which man lives in relationship and in balance with other living beings."[1] If one of these elements is altered, the quality of life worsens, as it can currently be seen.

Environmental conditions have been deteriorating year after year.[2] Global temperatures are rising progressively and are estimated to increase at least by 1.8°C by the end of this century. The balance between the cycles of rainfall and drought is altered: each phase of this cycle has been intensified.[2]

Ethiopia has experienced periods of drought lasting two years at a time, causing the death of thousands of living beings. If upper levels of evaporation increase, precipitation also increases; clouds, by retaining heat, promote the greenhouse effect. Because of the change in CO_2 levels, seasonal growth cycles of plants have also been altered, causing food instability and related deaths.

According to the World Health Organization (WHO), Italy is now—overall—a country with medium-high water stress.

Currently, the basin of the Po, Italy's longest river, has a water deficit of 61% compared to the previous years. In 2022, Irpinia, an area known for its vineyards and hazelnut groves, suffered a lack of rainfall which cut agricultural production by more than half.

---

[1] Italian Encyclopaedia Treccani, 1970.
[2] IPCC (Intergovernmental Panel on Climate Change), 1988.

Elsewhere, glacier melt at the poles has caused sea levels to rise by 20 to 50 centimeters. On July 3, 2022, because of high temperatures, a piece of glacier broke away from the Marmolada mountains resulting in deaths, injuries, and reports of missing people. Additionally, worsening climate conditions, have resulted in more intense rainfall and storms in other regions, with many areas subject to flooding. In November 2022, heavy rains caused a landslide on the island of Ischia in the Gulf of Naples that surged from Mount Epomeo toward the sea. The communi Casamicciola Terme was devastated; houses, shops, and vehicles were destroyed and twelve people died.

On December 23 of that year, temperatures in the US held most states below freezing. More than a million Americans remained without electricity. Twenty-eight victims of frostbite were confirmed.

There are various causes for this degradation.

In this regard, Renato Mazzoncini, in his book *Inversione a E* ("E turn")[3] claims that the first global trend that must be taken into consideration is demographic growth. According to estimates from the United Nations, the world population amounts to 7.9 billion people. In tens of thousands of years, we reached one billion people and in just 200 years we went from one billion to nearly eight billion. At this rate, the global population will reach 10.9 billion people by 2100 and only at that point growth will probably stop.

This excessive population growth has led to significant energy consumption. But, another factor to take into consideration is urbanism, or the migration of people from rural areas to cities, and with that a considerable increase in the use of vehicles powered by internal combustion. The development of mobility must, therefore, introduce the use of non-linear or renewable circular energy sources. In 2012, in Italy, it was estimated by the EEA (European Environment Agency)[4] that transport is responsible for

---

[3] Renato Mazzoncini, "E-turn," Egea Editions, September 2021. ISBN 978-88-238-3828-4.
[4] EEA European Environment Agency. Trends and projections: Limited Increase in EU Emissions in a Context of Recovery from the Pandemic and Energy Crisis.

approximately 24% of the total greenhouse gas emissions affecting the energy balance.

The theme of "green mobility," therefore, assumes a predominant role in current society, as well as in national, European, and global political agendas.

Since its 1947 inception, the Italian Constitution has protected the country's environment and inhabitants. Article 9 states among other things: "The Republic... safeguards the natural landscape and the historical and artistic heritage of the Nation."[5] Article 41 even specifies the following: "Private economic enterprise is free. It may not be carried out against the common good or in such a manner that could damage safety, liberty and human dignity."

The 1997, Kyoto Protocol brought attention to global warming. The 2016 Paris Agreement placed the focus on CO_2 emissions reduction to keep the rise in average global temperature below 2°C. The 2023 G7 Ministers' Meeting on Climate, Energy and Environment in Sapporo ordered a halt to plastic pollution.

Student movements around the world are protesting in an attempt to have their respective governments to adhere to these protocols, leading to a true generational and thought struggle, a clash between two different ways of seeing the world. On one side, there are the traditionally favored economic models based on profit and on perpetual growth, on the other side, there are supporters of sustainable economies. An entire generation of young people (i.e., Generation Z)—led by Swedish environmentalist, Greta Thunberg—protest by merely asking to live in a world without the likelihood of imminent environmental catastrophes.[6] The problem is, therefore, systemic and rooted in two key aspects. The first is the superficiality of the older generation and a lack of understanding regarding the natural interdependence of environmental elements; the second is the desire of many to change the economic system in favor of a new model of sustainability (and how it is manifested economically). This generational clash, in the recent

---

[5] From the Italian constitution: "The Republic promotes the development of culture, and scientific and technical research. It protects the landscape, the historical and artistic heritage of the nation. It protects the environment, biodiversity and ecosystems, also in the interests of future generations."
[6] Interview with Greta Thunberg, Swedish activist for sustainable development, December 14, 2022.

years, has been attenuated by a greater environmental awareness from the general public. Ultimately, the objective—which is gaining more and more support—is to decarbonize the production of energy, or more specifically, enabling human activities to continue without the use of fossil fuels.

FIGURE 1.1 Schematic summary of the forms of air pollution.

Figure 1.1 shows a schematic summary of the forms of air pollution: global and local pollution, caused by the excessive production of CO_2, CH_4, thin PM (2.5–10 micrometers), ultra-thin PM (10 manometers), noise, (which can exceed 10 decibels) and nitrous oxide N_2O.

Figure 1.2 shows the road map of decarbonization and the conversion of zero-emission mobility with the use of green hydrogen as it will be also explored in Chapter 4.

FIGURE 1.2 Road map of the conversion of mobility to zero emission using green hydrogen.

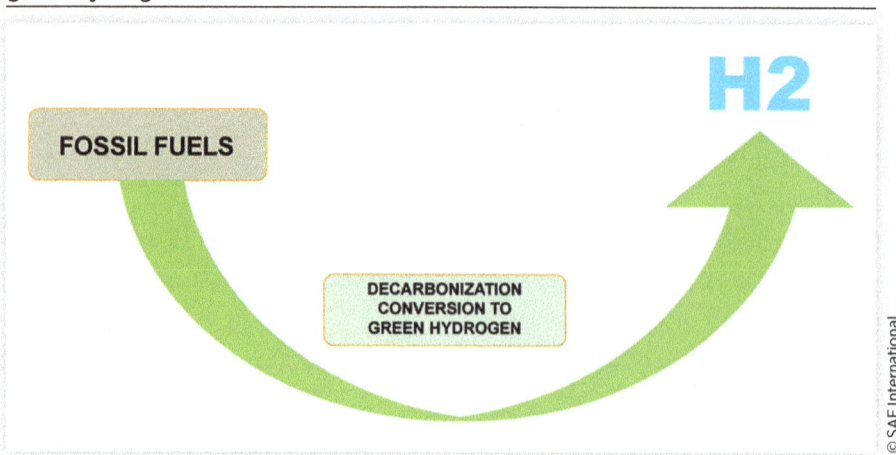

1.2. Causes of Environmental Degradation

1.2.1. The Greenhouse Effect

The main cause of global warming and pollution is the greenhouse effect.[7] As the sun's rays heat the surface of the Earth, they are reflected as infrared rays. Greenhouse gases, present in the atmosphere, retain the reflected infrared radiation, altering the Earth's temperature.

The main greenhouse gases are carbon dioxide (CO_2), methane (CH_4), water vapor (H_2O), carbonaceous particulates, particulate matter (PM), nitrous oxide (N_2O), nitrogen oxides (NOx), and the fully halogenated compounds of fluorine (CFC) and bromine (BFC). These substances derive from human activities, such as mobility, domestic heating, and energy production, as well as deforestation.

Recent studies, conducted in the laboratories of the CNR of Naples with the laser light scattering technique, have highlighted the presence of ultra-thin particles (i.e., 10^{-9} nanometers) in the exhaust of vehicles with endothermic engines[8] and also from industrial combustion (e.g., industrial burners, boilers, gas turbines, reactors), civil combustion (e.g., domestic boilers, fireplaces, wood ovens), industrial incinerators, and waste-to-energy plants. Furthermore, we cannot fail to consider, as Mariano Migliaccio rightly maintains, other types of particles, even very small and widespread ones, which are the result of the wear and tear continually present in the context of vehicle mobility. These have no relation with combustion phenomena and therefore cannot be easily eliminated unless urban mobility models are extraordinarily different from the current ones. Examples of this include particles from friction braking systems on vehicles (of lesser importance on EVs with partially electric braking), tires (significantly relevant compared to combustion particles of modern internal combustion

---

[7] IPCC (Intergovernmental Panel on Climate Change), 1988.
[8] CNR - STEMS - Institute of Sciences and Technologies for Sustainable Energy and Mobility.

engines), and road pavement, as well as industrial production activities involving stone and ferrous materials.

These sources release PM during their intended use. From some summary estimates, it can be stated that the particles produced by the exhaust of internal combustion engines in mass can reach a few percent (2–3%) of those produced by all other sources. Only as an order of magnitude it can be stated that percentages of 5–7% can be produced from vehicle braking systems, 15–20% from tires, and 15–20% from road wear. The rest of the particles are attributable to uncontrolled combustion phenomena (e.g., fireplaces, forest fires) and domestic and industrial sources mentioned previously.

These particles are extremely harmful to the respiratory system. Due to their small size, they remain in the atmosphere for long periods of time before falling to the ground. They absorb bacteria and viruses which, by entering the organism without being expelled, cause mutagenic and carcinogenic actions.[9] The production of nitrogen oxides is also, in part, attributable to natural events, such as lightning, which can occur several million times a day on Earth. Naturally, humans have little influence over some of these sources; however, there are those more directly connected to human activity that can be mitigated. It is precisely from the increase in temperature, exacerbated by the greenhouse effect, that humanity is faced with the accelerated glacial melt, drought, tropical storms, floods, and the extinction of many terrestrial and aquatic animal species.

1.3. Treatment of Municipal Solid Waste

Proper waste management is essential to protect the health of the environment and its inhabitants. It involves different phases (i.e., collection, transport, recycling, reuse, and disposal), and—in Italy—is controlled by European, national, and local regulations.

In September 2020, the Ministry for Environment, Land and Sea Protection, in collaboration with the Institute for Environmental

---

[9] Prisco Piscitelli, Matteo Rivezzi, Felice E. Corcione, and Alessandro Miani, "Air Pollution and Estimated Health Costs Related to Road Transportations of Goods in Italy: A First Healthcare Burden Assessment." International Journal of Environmental Research and Public Health. August 2019.

Protection and Research, established the National Program for Waste Management and defined the criteria and objectives that the regions, provinces, and local governments must comply with. One of the priorities of the program is the separation of different types of waste by citizens: organic wet fraction, dry fraction, glass, paper and cardboard, and plastic and metals. The correct disposal of separate waste leads to the recycling of differentiable waste, with the recovery of raw materials and energy, and less final product destined for landfills or waste-to-energy plants.

In 2008, as mayor of the city of Marigliano, I found myself managing the huge waste crisis in the province of Naples. Government commissioner De Gennaro decided to send 40,000 tons of unsorted waste from the city of Naples and its surroundings to my city. The citizens strongly opposed it because, despite recycling 70% of waste, they felt as if they were being punished. The bales of waste were deposited at a site in Boscofangone, home to the wastewater purifier of the Campania region.

My commitment was to protect the aquifers from the leachate that developed following meteor showers on the bales of waste containing the mixture of dry and wet organic fractions with geological barriers.

1.4. Landfills

Landfills are where unsorted waste is deposited. Such waste, solid or wet (e.g., industrial scraps, debris), cannot be recycled and is not sent to waste-to-energy plants for energy production.

According to the Italian legislation of January 13, 2003, in application of European legislation, three types of landfills are distinguished for inert, non-biodegradable materials; non-hazardous waste; and dangerous waste, such as ash and waste from incinerators. To limit harmful emissions, modern landfills must comply with environmental safety regulations and must be built in such a way as to isolate waste from the ground with geological barriers impervious to highly polluting leachate. Leachate is a liquid (e.g., rainwater) that permeates waste and carries soluble or insoluble waste solids with it. According to landfill regulations, leachate must be isolated from the aquifer and treated on site. Landfills must be subjected to repeated and constant checks according to the parameters of international regulations, from design to construction to disposal.

1.5. Waste-to-Energy Plants

Unlike incinerators which burn residual waste and send the exhaust gases directly into the atmosphere, waste-to-energy plants convert heat generated by the combustion of dry fraction into thermal energy. This heat is used by turbogenerators, which in turn produce electrical energy. Due to the poor quality of the waste introduced into the combustion chamber, the gaseous and particulate emissions at the exhaust are highly harmful if not appropriately treated.

The installation of the Acerra Waste-to-Energy power station,[10] one of the largest in Europe, was proposed in the 2000s. To evaluate the effects of exhaust emissions on the already highly degraded Marigliano-Acerrano territory, in 2008 it was established, on the indications of the bishop of Acerra, Mons. Rinaldi, the committee of experts from the academic and scientific world I was part of, as director of the CNR Motor Institute of Naples.[10] On March 22, 2010, the conference was held in the diocesan library of Acerra with the participation of:

- Professor Francesco Paolo Casavola, President Emeritus of the Constitutional Court of Italy
- Don Giovanni Rinaldi, Bishop of Acerra
- Dr. Tommaso Esposito, Mayor of Acerra
- Dr. Prisco Piscitelli of the Italian Society of Environmental Medicine
- Professor Luigi Fusco Girard, University of Naples Federico II
- Professor Aniello Montano, University of Salerno
- Professor Amalia Virzo De Santo, Federico II University of Naples

---

[10] Monsignor Giovanni Rinaldi, Bishop of Acerra, "The Incinerator of Acerra, Prehistory of Laudato Sì in the Times of Covid 19." Tipografia F. Capone, Editore of Acerra ISBN 9788895255323, 2021.

The conference—titled, "Defending the Environment to Defend Man and Society"—covered topics such as the philosophical foundations of environmental protection, waste management and sustainability, waste-to-energy plants and territory, quality monitoring, and effects on the health of living beings.[10]

Professor Casavola gave the talk "Man and the Environment in Benedict XVI Pope's Encyclical Caritas in Virtute," which states: "Nature is at our disposal not as a pile of rubbish scattered at random, but as a gift."[10]

1.6. Agriculture

Agriculture has always been of primary importance regarding human progress. Over the centuries, we have moved from subsistence farming to extensive agriculture based on large estates and crop rotation to industrialized, intensive agriculture aimed above all at promoting the economic development of the country concerned.

In the study of agriculture, three main subjects can be analyzed: the physical environment, the purposes and characteristics of the operators in the sector, and the economic and agricultural policies in play.

Intensive agriculture, as opposed to extensive agriculture, is intensive because it is increasingly based on the input of external energy into the system in the form of pesticides, mechanization, fertilizers, and genetic engineering.

The chemical composition of the soil and subsoil, therefore, is altered by non-biodegradable waste, wastewater, the excessive use of plant protection products, fertilizers, hydrocarbons, heavy metals, and organic solvents.

1.7. Transport

Transport has significant economic and social importance, as it promotes interpersonal relationships and the exchange of material goods and knowledge, moving people, animals, goods, and information from one place to another. It mainly uses vehicles with endothermic engines powered by liquid and/or gaseous fuels of fossil origin.

There are approximately 250 million cars in circulation in Europe with an average age of 12 years. It is estimated, as previously reported, that transport contributes to approximately 24% of air pollution (compared to the 76% related to other uses, such as civil and industrial).[11]

At a global level, the International Organization of Automobile Manufacturers estimated that 26,082,220 vehicles—21,407,962 cars and 4,674,258 commercial vehicles—were produced in the People's Republic of China in 2021, practically a third of the entire world production, although far from the 2017 record of 29,015,434 units assembled in the country.

China therefore represents the true nerve center of mobility. The second position goes to the US with 9,167,214 vehicles in total, and the third to Japan with 7,846,955 produced during 2021. With those two countries added to together, the number is approximately 12 million units less than China's production. Because of the expected ban on vehicles with internal combustion engines and the launch of EVs, as well as the low costs of labor and transport of electric traction components, automotive manufacturers want to manufacture in China. Thus, numerous joint ventures have been formed with Chinese manufacturing giants to produce new models ready to pervade the global market—all to the detriment of European domestic production.

1.8. Cost Analysis Between Diesel and Electric Cars

I tried the intellectually honest exercise of imagining a future with plug-in hybrid and battery electric vehicles (BEVs). With the context of a classic "home-work-home" routine today, I travel 100 kilometers a day on average, arriving and parking in the street outside my house.

I have a 10-year-old diesel car that achieves 20 kilometers per liter of diesel fuel. With the current price of diesel, this costs around €9.25 per day. If I had a plug-in hybrid or BEV and followed the

---

[11] Statistical Office of the European Union.

same route, I would arrive at the company car park with no possibility of recharging. After work, I would arrive home and go looking for a charging station (I do not have a garage). In the surrounding area, I have two options: the best one is a 10-minute walk from my house with only one charging station. The other one is 15 minutes from home, similarly with only one charging station. (Both are overseen by the same manager.) Taking advantage of the operator's best offer, I top up at €0.62 per kilowatt-hour. Imagining a consumption of 17 kilowatt-hours per kilometer, I spend around €10.58 per day. Diesel is better than electric. To me, this explains why the diesel car is more convenient and easier to use.

Every step is a choice, every step makes an imprint.

—*Luciano Ligabue*

2

European Union Proposals

2.1. European Union Proposals: Fit for 55

In 2008, the European Union had focused attention on climate change by proposing a 20% reduction in greenhouse gas emissions for 2030, a 20% increase in energy efficiency, and a 20% contribution of renewable sources in the energy mix.

At the end of 2019, member states once again sent their energy and climate plans for the period of 2021 to 2030 to the commission, so as to contribute to the mentioned climate objectives. On July 14, 2021, the commission then adopted the Fit for 55 climate packages,[1] which established the legislative proposals to achieve the decarbonization objectives by 2030. In particular, it proposed the reduction of greenhouse gas emissions by 55%, compared to the levels of 1990, to reach total decarbonization by 2050. Italy and Germany, with other countries and representatives of companies in the transport sector, did not fully agree. While they accept the reduction of polluting emissions, they argue that switching to 100% electric mobility is too much, as it would bring with it economic risks related to the loss of numerous jobs and the adaptation of production sites to new technologies. Europe's objectives have only been partially achieved. Emissions, from 1990 to 2019, were reduced by 23%.

---

[1] Fit for 55 - The EU plan for a green transition.

Therefore, I ask myself the following: if in 30 years emissions were reduced by only 23%, how would it be possible to reach the objective of reducing them by 55% by 2030? Industry experts are developing suitable solutions—some have talked about an energy revolution with reference to the date of July 14, 2021, while others, in disagreement, say that Fit for 55 seems only to be an "advertisement for a gym."

The subject is complex, confusing, and constantly evolving. In early March 2023 the production ban on internal combustion vehicles was postponed; the doubts of Italy, Germany, and car manufacturers were acknowledged.

On March 13, 2013, Thierry Breton, Commissioner for Internal Market of the European Union, referred to the 2035 ban on new fossil fuel vehicles and recommended manufacturers wait for further decisions from Brussels, according to which electric and endothermic power must go forward together.

2.2. New Proposals and Actions from Europe

The European Union energy and climate plans[2–5] for the period of 2021 to 2030 have been so much opposed from manufacturers and from the governments of member states that, as already mentioned, the decision to block the registration of vehicles with endothermic engines, set for 2030, has been postponed. The European Parliament has therefore approved the ban on the sale of vehicles with internal combustion engines, proposing the solution of electric and hybrid vehicles, as well as those powered by green hydrogen, biofuels, and/or synthetic fuels. EVs will reach maximum diffusion when the performance and economic advantages exceed those of vehicles with internal combustion engines.

---

[2] From the Italian constitution: "The Republic promotes the development of culture, and scientific and technical research. It protects the landscape, the historical and artistic heritage of the nation. It protects the environment, biodiversity and ecosystems, also in the interests of future generations."
[3] CNR - STEMS - Institute of Sciences and Technologies for Sustainable Energy and Mobility.
[4] July 2018. Visit to the Lawrence and Sandia National Laboratories in Livermore, California.
[5] European Commission: recent proposals to decarbonize the planet.

As the world is changing, we need to stay at the forefront of innovation to take advantage of this technological conversion. Car manufacturers will be able to achieve their objectives within the expected timetable, the fact that industry and trade reached the 55% of agreement. Ready package proposal demonstrates the European Union's determination to make progress toward climate neutrality and going green. Zero-emission mobility will help to slow climate change and avoid serious disruptions in various areas of society, such as environmental health, migration, food security, and the economy.

In 2026, as agreed, the European Commission will thoroughly evaluate the progress made to achieve the 100% emissions reduction targets. These objectives will take into consideration technological developments regarding plug-in hybrid technologies and the importance of a sustainable and socially fair transition. The agreement also provides for further provisions of the regulations, such as the reduction of the maximum limit on emission credits. Manufacturers receive such credits if they reduce CO_2 emissions from seven grams per kilometer per year in road transport to four grams per kilometer per year from 2030 to 2034.

The commission will develop a common European Union methodology to assess the full life cycle CO_2 emissions of cars and vans placed on the European Union market, as well as the fuels and energy consumed by these vehicles. Manufacturers can voluntarily report emissions produced by new vehicles they place on the market. Small manufactures will be able to comply with the legislation until the end of 2035. The provisional political agreement will then have to be formally adopted from the European Council and Parliament. The European standards on polluting emissions, created in the 1990s, have therefore limited the emissions of vehicles sold in European Union member states, achieving notable progress in reducing NOx and particulate emissions, from 0.75 grams per kilometer in 1995 (i.e., Euro 2) to 0.12 grams per kilometer in 2015 (i.e., Euro 5)—a reduction of more than 80% (Figure 2.1).

FIGURE 2.1 Evolution of European legislation from 1995 to 2015.

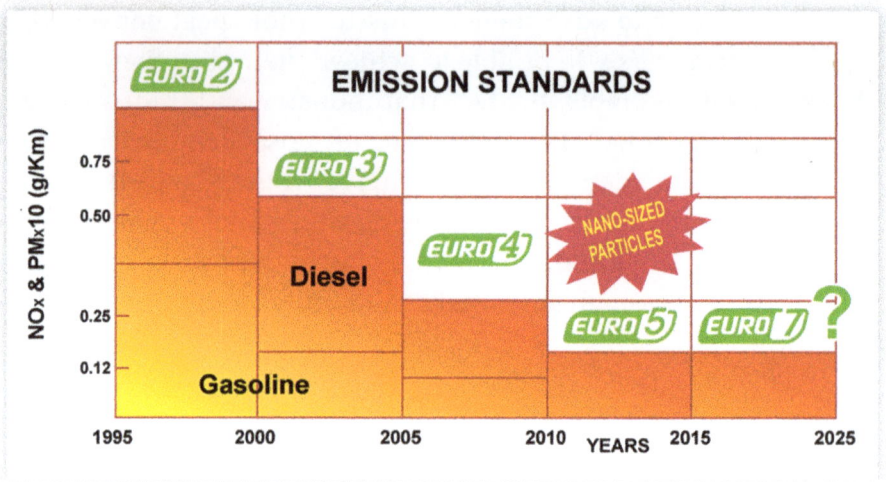

2.3. Euro 7 Regulation

Euro 7 standards set the specifications for each pollutant of vehicles on the road. The objective is to guarantee cleaner vehicles, including electric ones, thus encouraging their uptake.

In November 2022,[6] Europe announced the application of this regulation, which obliges manufacturers to further reduce NOx and particulate emissions starting July 1, 2025 for cars and July 1, 2027 for heavy vehicles.

Emissions will have to be reduced by a further 50%. NOx emissions will go from 90 milligrams per kilometer for gasoline-fueled cars to 120 milligrams per kilometer for diesels, with CO_2 emissions to 95 milligrams per kilometer for both types.

The legislation also establishes the reduction in emissions of fine dust and microplastics from brakes and tires for all vehicles, including EVs, and sets the minimum limit for the battery life of electric and plug-in hybrid cars.

---

[6] Fit for 55 - The EU plan for a green transition.

Why is the European Union tightening CO_2 emission rules for cars and vans? The answer is simple and obvious: there are numerous benefits for the public and the automotive industry. The population will benefit from cleaner air in cities where they normally live; manufacturers will have more time to adapt production sites to new technologies and, if they respect the required parameters, will be rewarded with less stringent CO_2 emissions targets.

We have to take care about nature as much as nature is taking care about us. Nature is very kind with us. And if you want to enjoy the gifts of nature and the promises of nature, we have to defer to nature and its needs, its rules, its norms.

—*Shimon Peres*

3

The Rebirth of the Planet in the Name of Renewable Energy

3.1. Introduction

Renewable energies are those available in nature—they do not pollute and do not run out. The most common ones are solar photovoltaic, new generation nuclear, nuclear fusion, wind, hydroelectric, geothermal, and biomass.

Photovoltaic is produced by solar panels converting the sun's energy into electricity. Until 2007, it was an experimental technology; with the incentive of the energy bill, its use grew exponentially until it reached 23 gigawatts in 2021 for a total of 987,000 systems, installed mainly in southern Italy.

It was precisely in that year that I designed and built a 500-kilowatt system in Vitigliano, which is a fraction of Santa Cesarea Terme, in the Province of Lecce.

For wind, turbines convert wind energy into electricity. As it is not a continuous resource, it is necessary to install more systems to produce more energy. Numerous turbines are visible along the A16 motorway between Naples and Canosa. In Taranto, the construction of an offshore wind farm, with turbogenerators on floating platforms, is at an advanced stage.

Hydroelectric energy is produced by the force of moving water in hydroelectric power plants. It is the most widespread renewable energy application in Italy, as it makes up 40% of total electricity production. Professor Cristiano Corsini of Roma Tre University stated in November 2021, "If we want to extract more energy, the only path to follow can come from the creation of reversible systems that store energy at night and return it during the day."

Geothermal uses heat coming from the deepest layers of the Earth and ambient solar heat to produce electricity through turbogenerators.

Energy produced from biomass uses organic materials such as wood, agricultural waste, and waste to produce energy.

In 2019, the International Renewable Energy Agency estimated that such energies cover three-quarters of global energy and "promotes the widespread adoption and sustainable use of all forms of renewable energy, in pursuit of sustainable development, access to energy, energy security, economic growth and a low-carbon perspective."

The conversion to zero emissions requires financial, cultural, and experimental investments, as well as the involvement of national governments, public and private research centers, universities, and companies in the sector.

3.2. Solutions and Remedies

In March 2023, the European Commission "opened" to negotiations with Italy on biofuels and with Germany on synthetic fuels. It suggests three possible solutions to adopt:

- Biofuels and synthetic fuels; for internal combustion engines already in circulation.
- Green hydrogen; for vehicles with an internal combustion.
- Electric fuel cell engines.

In agriculture, two methods are suggested to minimize soil, air, and water pollution: organic and biodynamic options, both based on biodiversity and crop rotation by using self-produced fertilization preparations. These routes are determined to take care of the processes that are linked to the ground on one side and those, such as sunlight, on the other, keeping them in harmony. Cultivation, thus, "lives" in the balance between these two forces.

At Vigna Maurisi, I use the biodynamic method, certified by the "Suolo e Salute" organic certification body. The vineyard, located far from sources of pollution, is always grassed because highly carcinogenic herbicides (e.g., glyphosates) are not used. With the autumn sowing of various flowing species, which are then planted in the spring, biodiversity is increased, making the vineyard like a spontaneous natural ecosystem. In spring, the grass is tall, full of flowers, with deep and vigorous roots; it is the true creator of the soil structure. On top of that, proceed with shredding and inter-row brush cutting; the burial of the organic substance quickly transforms into humus. To combat fungal growth, copper and sulphur are used. Natural manure also helps the formation of humus in the soil, adding to soil fertility. The water used, from rain or spring sources, becomes the living carrier of the "information" on the vines (Figure 3.1).

FIGURE 3.1 Vigna Maurisi is cultivated with the biodynamic method.

3.3. Biofuels

The use of alternative fuels of vegetable or synthetic origin is proposed to save internal combustion engines. In this regard, Mariano Migliaccio claims, "For some years, alternative fuels have been developed, so to replace fossil fuels in many industrial applications, first of all those ones linked to energy production and transport."

These fuels are characterized by low carbon emissions and are designed to replace fossil fuels in heat engines without any substantial modification.

Cummins has placed diesel generator sets on the market that use hydrotreated vegetable oil which has the characteristic of being completely renewable. The company also produces diesel thermal engines that can be powered by biodiesel blends obtained by mixing normal diesel and renewable vegetable oil.

Nowadays the use of biogas (consisting mainly of methane and renewable CO_2), obtained by the treatment of urban or industrial waste in anaerobic production plants, is also becoming increasingly widespread. These can directly fuel industrial engines for energy production. Recently, the technique of producing biomethane from the same biogas production plants has been spreading, adding appropriate downstream treatments for the separation of noncombustible products. These can have a high production purity with the advantage of reducing the consequent emissions.

Naturally, the production of biomethane from the recycling of urban waste represents a chance to produce fuel for the metropolitan transportation fleets, a technique that is also encouraged by numerous government initiatives in different countries around the world.

For that purpose, the company ENI has patented the Ecofining™ system, which processes biomass into high quality biofuels (e.g., hydrotreated vegetable oil). It uses the trans-esterification process, by which incoming triglycerides are treated with methanol to obtain a product whose characteristics strongly depend on the type of raw material.

The process deals with different types of organic products such as waste animal and vegetable fats from the food industry and used cooking oils. The company declares that it has developed these technologies in its laboratories and research centers in collaboration with Honeywell UOP. The two companies have converted two conventional refineries, the one of Venice and the one of Gela, into biorefineries, contributing to the relaunch of the refining sector. Traditional biofuels derived from crops are often in competition with food use, while advanced biofuels treat waste or crops such as straw, raw glycerine, agricultural and forestry cuttings, palm oils, rapeseed, and organic waste.

With this technology, a high-quality biofuel is obtained without oxygenated components and with a high cetane number which can be used in current diesel engines.

3.4. Synthetic Fuels and "E-Fuels"

Synthetic fuel, also known as electronic fuel or "e-fuel," includes liquid fuels obtained from coal, natural gas, oil shale, biomass, or environmental CO_2 with the addition of hydrogen. The name can also refer to fuels derived from solids, such as plastic or rubber waste, following separate waste collection. The energy-intensive process was first developed in 1925 by Franz Fischer and Hans Tropsch in Germany and can be powered by renewable electricity.

Synthetic fuels are an economically viable alternative to oil. They are a useful source to maintain the economy because they can be used directly in internal combustion engines and eliminate the costs of adapting production sites of the industries in the sector. However, there are still many questions surrounding production costs.

3.5. **Circular Economy**

A circular economy is a self-regenerating economic system that guarantees its ecological sustainability.[1]

The principle of the circular economy is not entirely new. For different reasons (mainly linked to economic constraints), it was already in use in the 1950s and 1960s. In peasant families, nothing was wasted—everything was used and reused several times. Clothes, once worn, were turned inside out, thus acquiring a new life and value. People foraged and collected various edible wild plants and herbs and preserved them for the winter. With the advent of the economic boom, the standard of living improved and reuse was abandoned in favor of disposable products, resulting in wasted energy and increased waste and pollution.

Today, the concept of a circular economy is increasingly replacing the linear economic model. As the latest statistical data shows that more than 2.5 billion tonnes of waste are produced every year, the circular model has become one of the main objectives of the European Union, which is updating legislation on waste management to facilitate the transition toward a green economy. In fact, the action plan for a new circular economy with the use of more sustainable products to reduce waste was presented as the "European Green Deal."

What is the circular economy exactly? What is the difference from the linear model? Why are companies pressing for this change?

The linear economic model, unlike the circular one, is based on the typical "extract, produce, use, and discard" involving the extraction of raw materials, mass production and consumption, and the disposal of waste at the end of product life. It depends on the availability of large quantities of low-cost materials and energy, which are easily available.

---

[1] European Commission: Recent proposals to decarbonize the planet.

Replanet Magazine,[2] a well-known marketing magazine from the Veneto Region, defines the circular economy as "a model of production and consumption which involves the sharing, lending, reusing, repairing, reconditioning, and recycling of materials and products existing for as long as possible." In this way, the life cycle of products is extended, and the waste cycle is reduced. When the product ends its function, it does not die: the materials that comprise it are reused several times in the economic cycle, generating more value.

There are many benefits of the circular economy. By reusing products, consumption of natural resources is reduced and ecological health is preserved. Greenhouse gas emissions, energy consumption, and the use of raw materials decrease. In this regard, we should mention the collection and reuse of waste electrical and electronic equipment (WEEE) which recover precious metals such as gold, platinum, and silver.

Waste, as a resource, therefore, has a positive economic impact. Soon, competition will increase among companies using recycled products.

---

[2] Renato Mazzoncini, "E-turn," Egea Editions, September 2021. ISBN 978-88-238-3828-4.

The most common elements in the universe are hydrogen and stupidity.

—Harlan Ellison

4

Hydrogen as an Energy Carrier of the Future

4.1. Hydrogen

Hydrogen is a fuel with a large energy density. In the free state, it is found in the form of diatomic gas (H_2).[1,2] It is colorless, odorless, tasteless, and highly flammable. Among the many elements making up matter, it is the lightest and most abundant. It makes up almost 90% of the visible mass of the universe. In its gaseous form, made up of a simple two-atom molecule, it burns in a similar way to methane or other gases in an oxygen-rich atmosphere (like that of Earth).[3]

Among conventional fuels, it is the one with the highest energy content per unit of weight—three times higher than gasoline. It is used as a raw material in industrial production processes, ammonia, fertilizer, and oil refining applications. It is produced almost entirely from fossil fuels through the gasification of coal or the steam reforming of natural gas. These processes, which do not use CO_2 capture systems, are polluting. It is therefore necessary to resort to other "cleaner" production methods, including water electrolysis, which uses electricity produced from renewable energy sources and nuclear power. Hydrogen, with low CO_2 emissions, can be used in various applications, such as fuel cells for

---
[1] Hydrogen according to Enel S.P.A.
[2] Sustainable Energy Research Institute "Sotacarbo S.P.A." of Carbonia (SU- Sardinia - Italy).
[3] Fit for 55 - The EU plan for a green transition.

long-distance air, land, and marine transport; energy storage and power buffering to support greater penetration of renewable energy in electrical and chemical networks; and internal combustion engines and gas turbines. It can also be blended with natural gas and added to pipelines to provide cleaner fuels for civil and industrial applications as a complement to electrification.

Hydrogen's mass energy content makes it suitable as a transport fuel. On a volumetric basis, as the energy content is relatively low, larger volumes should be used to meet the same energy demand as those achieved with gasoline and natural gas. The production of clean hydrogen can help solve the energy challenge of our planet. However, it has just one problem: producing it is not simple. Conversion to clean hydrogen requires adequate financial and regulatory support for the supply of fossil fuels, water resources, storage, and infrastructure. Still, the development of technologies to obtain it in a clean way, with the help of renewable sources, opens up a new green future.

Furthermore, hydrogen is the fuel that powers the nuclear fusion of the sun and stars. The fusion of hydrogen atoms produces the renewable thermal energy that the Earth receives from the sun every day in the form of radiation. In fuel cells, combined with oxygen, hydrogen produces electricity and water vapor. Electrification, through renewable energy with hydrogen and in combustion systems, will be the main way to decarbonization. However, some end uses are still difficult to decarbonize through a direct electrification process. To overcome this, the use of green hydrogen is essential. In transport the use of green hydrogen, as described below, will still require many years of research and experimentation.

In fact, energy policy analysts predict a broader role for hydrogen. As it represents up to 30% of the energy supply in several sectors, demand will increase more than fivefold by 2050.[4]

---

[4] https://www.irena.org/.

4.2. Different Types of Hydrogen

Hydrogen, depending on the way it is produced, takes a different name with colors indicating a specific distinction, such as green, grey, and blue. These names are symbolic and refer to different gas production processes. Produced with electrolyzers powered by renewable electricity (e.g., wind, photovoltaic, geothermal, nuclear), "green hydrogen" has environmentally friendly properties.[5]

It has been estimated that around 3% of the world energy consumption will be used to produce hydrogen. Although less than 0.01% of this hydrogen, around 1000 tonnes per year, is used as an energy carrier, there is growing interest in its ability to reduce greenhouse gas emissions as a clean energy carrier. Green hydrogen burns without producing harmful exhaust gases; here the only waste product is water vapor.

"Grey hydrogen" is produced by the steam reforming of methane or by the gasification of coal with high CO_2 emissions. Generated during the processing of fossil fuels whereby natural gas is converted into hydrogen and CO_2, it is by far the most common production method. While CO_2 is released into the environment, hydrogen is reused. Production requires extreme heat and sometimes appears as a waste product from the industrial sector. Therefore, it has a negative emissions impact because it produces CO_2. It follows that grey hydrogen occupies the last place in the ranking of the environmental compatibility of production processes.

"Blue hydrogen" is produced by the same methods as grey hydrogen, but involves CO_2 capture systems. Hydrogen, produced with electrolyzers powered by nuclear electricity, is called "pink hydrogen." When it is produced exclusively by solar energy, it is called "yellow hydrogen."

---

[5] Sustainable Energy Research Institute "Sotacarbo S.P.A." of Carbonia (SU- Sardinia - Italy).

4.3. Hydrogen from Ammonia

The production of hydrogen from ammonia (NH_3)—as per conversations with Professor Roberto Gentili of the University of Pisa[6]—requires high temperatures (i.e., 500 to 600° C) and is achieved with thermal energy produced by fuel that generates harmful emissions.

In November 2022, researchers at Rice University in Houston, Texas, created a revolutionary photocatalyst which, at ambient temperatures and pressures, uses light-emitting diodes (LEDs) to convert ammonia into hydrogen, accelerating the conversion reactions. Unfortunately, the researchers did not report the energy consumption of the LED lamp.

The application still requires development before being used at service stations on motorways and on state and provincial roads. Ammonia could be a better energy carrier than hydrogen due to its simpler management, but it involves costs related to produce it and split it into hydrogen. Additionally, the yields still need to be evaluated, with particular reference to the quantity of energy necessary for the LED lamps. Considering what has been said, ammonia could, however, be a valid alternative regarding the difficulties of storing and transporting hydrogen.

The decarbonization of transportation is a necessary path to reduce our impact on the climate. But it is still difficult to imagine what the future of cars will be when we finally abandon fossil fuels. Will we have an all-electric fleet of cars? Hydrogen engines? Some imagine a different alternative: ammonia, a fuel that does not emit carbon dioxide and does not require batteries, have pollutants to produce and dispose of, or use dangerous pressurized tanks (as in the case of hydrogen). In China, they have already announced the development of the first ammonia engine for cars. The work was a collaboration with the automotive giant Toyota, which seems to have decided to focus everything on the development of ammonia vehicles with the belief that they could mark the end of the nascent electricity market. The question, then, arises spontaneously: are ammonia engines really a promising alternative to decarbonize the transportation sector? We will see.

---

[6] https://www.ammoniaenergy.org/.

4.4. Ammonia Engine

There are several ways to use ammonia as a fuel. Ammonia is a molecule made up of three hydrogen atoms and one nitrogen atom. It does not contain carbon, and for this reason, its use does not directly generate CO_2 emissions. The high hydrogen content makes it possible to use it as a method of storing and transporting this element more easily than its pure form (which requires pressurized tanks at 700 atmospheres). One possibility, therefore, is to exploit ammonia for its greater ease of transport and storage, and then to break it down into hydrogen as fuel.

Alternatively, it is possible to exploit ammonia as a base to produce a fuel to use in a traditional internal combustion engine, employing other substances (e.g., diesel, hydrogen) to simplify the combustion process. However, ammonia alone is particularly complex. Finally, it is possible to develop a combustion engine powered by pure ammonia. While this is a more demanding engineering challenge, it would guarantee the elimination of CO_2 production, which in the previous case, even if drastically reduced, it would still be present.

4.5. The Announcement of Toyota

On November 15, 2023, Toyota announced the launch of a vehicle with a hydrogen internal combustion engine.

Will EVs be indispensable for the reduction of vehicle emissions? According to Toyota, they will not. And, in the future, there will not even be a need to stop the production of the internal combustion engine. In fact, a possible alternative could be to convert some mechanical engines to hydrogen.

Hydrogen combustion is an alternative method to make efficient engines from an environmental point of view. These types of engines have already been prototyped multiple times by the company and is being tested by the Japanese giant, on some GR Yaris and Corolla models. It is now planning to export on a "large" scale on the Land Cruiser off-road vehicle, after testing the technology in Australia.

4.6. Is it the Ammonia Era?

Leaving aside the engineering aspects linked to the creation of an ammonia engine (which are difficult to evaluate when talking about prototypes protected by industrial secrecy), there are several problems to overcome to transform this substance into the "gasoline of the future." Starting from its production, which is now linked to hydrogen, ammonia is normally obtained as a byproduct of hydrocarbons in an extremely polluting process. Green ammonia is therefore closely linked to green hydrogen, an expanding but completely minor sector. Imagining to use it as fuel for private cars requires estimating economic variables: if it is not possible to guarantee an expense comparable to what motorists now incur with gasoline or EVs, ammonia cars are unlikely to gain significant market share.

Ammonia is also an extremely toxic substance, which can be dangerous in the event of direct exposure (e.g., in the event of an accident) and it requires special precautions to ensure that it will not damage the engine or the tank because of its corrosive properties. The exhaust gases produced by its combustion also represent a problem. Although they do not produce CO_2 or other greenhouse gases dangerous to the climate, they emit large quantities of nitrogen, which can be harmful to human, animal, and plant health. With the right precautions, it is possible to limit nitrogen emissions (this is done in diesel engines). However, it is still a problem to keep in mind, especially after precedents such as Volkswagen's "Dieselgate" scandal when it was discovered that the vehicles of the German company produced NOx up to 40 times more than declared.

To find out if ammonia is really destined to undermine EVs, all that we have to do is to wait. It has strengths and weaknesses, like all alternatives to fossil fuels. But, of course, the trust that an automotive giant like Toyota seems to place in this technology can only be a plus.

4.7. Energy Revolution

Scientists predict that solar energy will dominate our electric grids by 2040.

Recent studies on renewable energy have highlighted that, for many years, climate scientists have been calling for a switch to alternative energy. This appeal came up against resistance stemming from political, economic, and practical perspectives. Recent research published in Nature uses a numerical model to indicate the point where solar and wind energy will be cheaper than fossil fuel.[7]

4.8. The Cost of Solar

According to Forbes, to install residential solar panels costs about $16,000 on average, depending on the type of panels and installation location.[8] However, this is a one-time expense, and many homeowners get their money back in 9 to 12 years.

Some local incentives can also offset this cost, making it a better choice than fossil fuels. In some states, energy companies will pay for the energy produced by privately owned panels. Some states also offer a tax credit for their installation. This credit is in addition to the federal tax credit for installing solar panels. In 2023, this credit was equal to 26% of an individual's federal income tax. These incentives will continue until at least 2035. The increasing trend of solar depends on the financial stability of the country.

Forbes also illustrates the rapid growth of solar energy that we have already seen. The cost of solar energy is decreasing as technology advances. This, along with the rising cost of fossil fuels, is pushing people toward solar energy.

This trend is particularly widespread in rich countries, which have the liquidity necessary to invest in the initial cost of solar energy. Developing countries will probably continue relying on fossil fuels, as that is the system in place now.

---

[7] https://www.snexplores.org/article/explainer-decarbonization-carbon-dioxide-climate.
[8] https://www.forbes.com/home-improvement/solar/are-solar-panels-worth-it/.

4.9. Progress of Nuclear Fusion

On December 21, 2022, fourth generation nuclear reactors were announced. A new generation of experimental or demonstration reactors based on nuclear fission, they are more efficient and safer than the ones of previous generations, producing less waste and at lower costs, making the use of fuel much more efficient. Around the same time, researchers at Livermore National Laboratory in the US announced a new method of nuclear fusion. This method can produce large quantities of clean energy with an energy gain of approximately 1.5 times the quantity used to trigger the reaction. It is a fundamental breakthrough that could completely change the way we produce clean electricity, but it will take decades of research and development before we harness these reactions.

If scientists confirm the results, it will be possible to have clean energy in large quantities to produce green hydrogen without resorting to climate-altering sources such as coal, oil, and natural gas.

4.10. New Generation: Waste to Energy Plants

A waste-to-energy plant is a plant that burns the dry, non-recyclable fraction of waste to generate electricity via turbogenerators. Newer generation units minimize harmful NOx emissions by using ammonia to treat the fumes in the combustion chamber. The exhaust fumes are purified and filtered to retain heavy metals, dioxins, and furans and then pass through sleeve filters to remove fine dust.

Researchers from the Polytechnic University of Milan, Polytechnical University of Turin, and the University of Trento have been experimenting with these processes for some years. The results of the research[9] state that the harmful effects on health "are attributable to old generation systems" used before the 1990s. "A recently designed plant emits relatively modest quantities of pollutants, it is not clear if it poses a real and substantial risk to health."

---

[9] Statistical Office of the European Union.

One of the best new generation waste-to-energy plants was inaugurated in 2017 in Amager near Copenhagen, Denmark.

A further innovation could, in my opinion, consist of directly using the hydrogen produced by an electrolyzer downstream of the plant itself (Figure 4.1). In this way, the use of ammonia would be eliminated with the combustion temperature thus avoiding the formation of dioxins and furans.

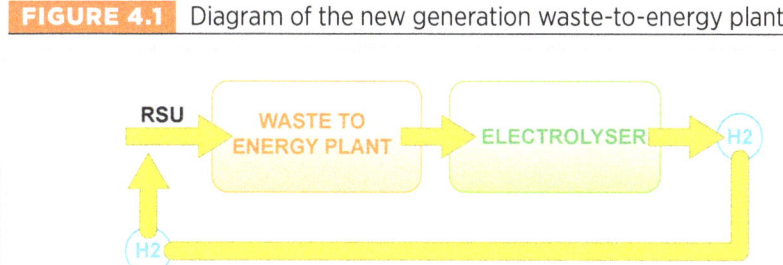

FIGURE 4.1 Diagram of the new generation waste-to-energy plant.

4.11. Anaerobic Digestion of the Wet Fraction of Waste

Anaerobic digestion of the organic fraction, or "FORSU," is a biological process that occurs with the absence of oxygen. Through this process, organic substance is broken down naturally by bacteria and turned into biogas and a sludge called "digestate" or compost, which is then used in organic and biodynamic agriculture. Biogas can be used for civil uses, transport, or for the production of electricity and hydrogen. From biogas it is possible to directly obtain biomethane, which is essentially a nearly pure methane. It has the advantage of being produced through the waste cycle, and, when used, it will produce the same CO_2 that has been absorbed by plants and substances that constitute organic fraction of municipal solid waste.

Some studies suggest that at least 10% of the fuel demand of the global transportation sector can be satisfied through the production of biomethane, thus contributing to the reduction of global CO_2 emissions.

4.12. Storage and Transport of Hydrogen

Hydrogen storage is fundamental in the development of the green economy. It can be stored on an industrial scale and recovered as a backup energy source when needed. For example, in processes requiring high temperatures—such as those adopted by the steel industry, in refineries, or in fertilizer production—stored hydrogen can be used as a feedstock for industrial applications. Furthermore, when properly mixed with natural gas, it can be transported through existing gas pipelines. Hydrogen storage research aims to achieve volume reduction for mobile applications. Such research is important to store the energy produced by photovoltaic and wind power. The main difficulties in using hydrogen as a storage system are related to complexity and production costs.

In these mentioned uses, it is important to highlight that, in all phases of production and storage, hydrogen does not generate CO_2 emissions or other forms of emissions harmful to humans or the environment.

Hydrogen can be stored and transported as a gas through heavy, high- pressure (i.e., 350 to 700 bar) metal tanks. A valid solution to improve this process is innovative composite tanks that combine high mechanical resistance with a lower weight.

Three types of hydrogen storage materials are being studied, including those using adsorption to store hydrogen on the surface of the material, those using absorption to store hydrogen within matter, and hydrides—chemical compounds where hydrogen is combined with another element using a combination of solid and liquid materials.

In 2018, the Mazda Motor Corporation developed a vast research program on the storage of hydrogen in metal hydrides.

> Electrification of mobility.
> An opportunity to start again.
> —*National Interuniversity Consortium on the Electrification of Mobility*

4.13. Electrification of Mobility

The electrification of mobility involves the application of electrical and electronic devices to promote the environmental, economic, and social sustainability of transport systems. These technologies concern green hydrogen production systems with electrolyzers, fuel cells for mobility, high-pressure gaseous hydrogen storage, and suitable refueling systems along roads and motorways.

According to transport industry leaders, it will still take 30 years before we will see hydrogen EVs on our roads. Only with a serious commitment from our politicians and the industry will the 2050 limit on green mobility be respected. The limitations are mainly related to the costs for green conversion, which are yet to be estimated.[10] When we carry out calculations about the average cost of EV charging services, we must consider two main items: the cost of the supply of electricity and the cost of the infrastructure necessary to carry out the charging. The impact that this new demand for energy may have on the electricity grids and low and medium voltage distribution networks (i.e., where the charging infrastructure will take place) must also be carefully evaluated. In this regard, it is worth remembering that the costs for the development, operation, and maintenance of electricity networks relate to the users of the electricity system.

4.14. Hydrogen Technology: Current Trends and Controversies

Figure 4.2 shows the diagram of a proton exchange membrane (PEM) type fuel cell. In it, gaseous hydrogen is sent to an anode covered with a cadmium catalyst. The hydrogen loses two electrons, giving rise to the electric current. Protons passing through the PEM encounter oxygen and form water vapor. The electrical energy, appropriately treated with an inverter and management unit, sends power to the vehicle's electric traction motor.

---

[10] IPCC—Intergovernmental Panel on Climate Change.

FIGURE 4.2 Hydrogen fuel cell.[10]

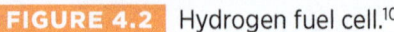

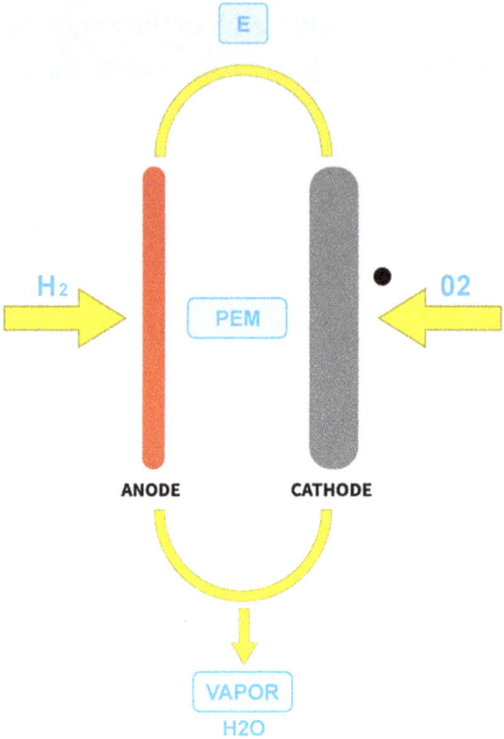

There are numerous advantages in the fuel cell:[11] it has high efficiency regardless of the size of the system, high energy density, and a relatively low weight and volume. Modularity (i.e., the possibility of combining multiple fundamental units) theoretically allows for power ranging from thousands to millions of watts.

In large systems, it is possible to further improve efficiency by recovering the heat produced during the reaction. Fuel cells have been used in space and military applications, in submarines and EVs, but because of safety issues related to the presence of hydrogen, they have not been used in other sectors yet.

The CEO of IVECO Fiat Group, Gerrit Marx, claimed, "What about e-fuel? Fuel only for the rich, with the electric car you risk having mobility only for the rich." He also directs heavy criticism

---

[11] Felice E. Corcione, "Let's Set Hydrogen in Motion." New Energy Sources, New Technologies, New Use and Diffusion Programs to Reduce Pollution, Conference at the Unione club in Marigliano, April 11, 2003.

of the Euro 7 standards, which I quote in full: "The decision to establish the Euro 7 standard for light commercial vehicles and trucks is stupid, because these are impossible times and in contradiction with the choice to focus on electric. This is the result of a European policy in the hands of people who do not know the automotive industry. They can ask us to invest in Euro 7 by reducing emissions from thermal engines or, alternatively, to invest in electric motors, but not to invest in both paths simultaneously in the next 10 years. Whoever decided on the new Euro 7 regulation simply created a working group with industry experts, who are also the suppliers of the companies and have every interest in making the standard difficult to achieve. They create work for themselves."

In response to this, Giuseppe Police maintains: "I absolutely agree with what the IVECO representative states. The choice of the European Union, in my opinion, is a rush forward which has no bearing on the world climate and which risks destroying our economy. I believe that the energy transition should contemplate the best possible use of an energy mix, not excluding fossil fuels and biofuels with the so-called renewables. Moreover, in my opinion, hydrogen is now expensive and difficult to distribute, unless it is mixed and distributed with methane in houses and cars. Aside from the costs, the transportation of hydrogen on board mobile vehicles is dangerous and not very acceptable to ordinary users. The use of hydrogen will probably become economically convenient when we have energy from nuclear fusion. Until then, the little hydrogen available should be used to form the so-called 'hitane,' (hydrogen and methane). Lately, I believe that the energy transition, as it has been proposed, is the swan song of Europe. The solution for the future is surely nuclear fusion with large quantities of electricity to produce hydrogen for civil, industrial, and transport uses, but everything still needs to be done."[12]

I perfectly agree with what was stated by Peppino Police and Mariano Migliaccio, who argue that the solution for the future is nuclear fusion, which can provide large quantities of electricity for the production of green hydrogen.

---

[12] Mariano Migliaccio, Marianna Migliaccio and Alessio De Stefano, "High performance gears (HSG)." Liguori Editore, March 2022.

Of a different opinion, Elon Musk, self-proclaimed "Technoking" of Tesla, at a conference on electric cars held in Germany in 2013, declared that, "hydrogen mobility is suitable for rockets, but not for cars," and the best fuel cell system is currently lithium-ion batteries.[13] In 2015, he reiterated that "hydrogen mobility is incredibly dumb,"[14] and then added, at the Automotive World News Congress in Detroit, that "If you're going to pick an energy storage mechanism, hydrogen is just an extremely dumb one to pick."[15] It should be emphasized that discrediting hydrogen mobility is in the full interest of Tesla, an empire based on electric mobility.

Not everyone, in fact, agrees with this thesis.

Giants such as Cummins Inc. in US; Mazda, Honda, Hyundai, and Toyota in Japan; the Stellantis Hordain Group, BMW, and Roll Royce in Europe; have already invested heavily in this technology. Costa Crociere has also ordered three liquid hydrogen fuel cell cruise ships from Fincantieri.

On May 10, 2023, the largest electrolyzer in the world with solid oxide electrolyzer cell technology began operating. It was created and installed by Bloom Energy—an American company also present in Italy through partnerships with various Italian companies, including Simplify and Cefla—at NASA Ames Research Canter in Mountain View, California. With just four megawatts of installed electrical power, 2.4 tons of hydrogen were produced per day with an overall efficiency greater than 80%. In Europe, German Chancellor, Olaf Scholz, and the President of the European Commission, Ursula Von Der Leyen, stated, "We are in a phase of constructive dialogue to guarantee the use of hydrogen in internal combustion engines. Germany, estimating that self-produced green hydrogen can only cover 30%, has begun to formalize green hydrogen supply contracts with foreign countries to cover national needs.[16]"

---

[13] https://venturebeat.com/business/tesla-spacex-ceo-elon-musk-fuel-cells-are-so-bullshit-and-hes-right/.

[14] https://dailycaller.com/2015/02/02/watch-elon-musk-explain-why-hydrogen-is-an-incredibly-dumb-alternative-fuel/.

[15] https://www.mlive.com/auto/2015/01/teslas_elon_musk_hydrogen_fuel.html#:~:text=%22If%20you're%20going%20to,in%20the%20next%20few%20years.

[16] https://www.cummins.com/news/2023/06/09/can-engine-run-hydrogen.

In the US, Cummins has estimated that "the conversion of medium and heavy trucks to fuel cells and/or green hydrogen would allow the elimination of around a quarter of all greenhouse gas emissions in the transport sector."[17]

With the Mirai fuel cell car, Toyota guarantees the performance of a true electric car, with a power of 182 horsepower and around 650 kilometers of autonomy. On February 17, 2023, Hyundai of South Korea presented the Hyundai Nexo hydrogen fuel cell SUV. Still, the list of fuel cell cars does not end here.

In Australia, researchers at the University of New South Wales have successfully transformed a diesel engine into a hydrogen engine. In 18 months, they developed a dual-fuel injection system which uses 90% hydrogen as fuel. Retrofitting existing diesel engines enables quick switching to a cleaner combustion system while minimizing development costs for companies. Generally, many companies in the transport sector are successfully directing production toward fuel cell vehicles and hydrogen internal combustion engines.

In 2014, the first plant in Italy for the production, storage, and distribution of green hydrogen (i.e., generated with the sole use of renewable energy sources) was inaugurated in southern Bolzano. Built by Autostrada del Brennero, the plant is managed by the Italian Institute of Technological of Genoa and is capable of producing 180 normal cubic meters of hydrogen per hour.

The topic of hydrogen is increasingly current, so much that the Department of Energy Technologies and Renewable Sources of the Italian National Agency for New Technologies, Energy and Sustainable Economic Development (ENEA), with the Italian Association of Chemical Engineering and the Department of Chemical Engineering, Materials, and Environment of La Sapienza, continue a "Hydrogen Summer School" on hydrogen technologies. Expert speakers from the academic and industrial world, doctoral students, researchers, and young professionals working in the sector continue to participate.

---

[17] Ibid.

The objective of the school is to offer a broad and complete vision of the supply chain, analyzing opportunities and potential and critical issues. It also addresses the role of hydrogen as an energy carrier and enabling factor for decarbonization and the energy transition. These initiatives confirm the beneficial change in mentality that can lead us toward a green economy by 2050.

4.15. Low Emission Cars

According to European plans to reduce CO_2 emissions, vehicles with new technologies must be marketed when they are inexpensive and accessible to everyone. Companies in the sector are experimenting with various fuel cell and hydrogen vehicle solutions, with the following types already available: full hybrids; plug-in hybrids; mild hybrids; full battery electric; and hydrogen fuel cell.

In the full hybrid models, both thermal engines and electric motors work to drive the vehicle. The supporting combustion engine significantly increases the vehicle's range.

Plug-in hybrids—without using their combustion engines—have a range of around 50 kilometers. They can connect to electrical sources or regenerate electrical charge through braking; however, charging is a slow process either way. To overcome this, gas stations should be equipped with spare, fully charged battery packs that can be conveniently swapped and traded for the empty batteries. This solution requires many years of research and development before it is possible to change the battery pack in national and European service stations.

Mild hybrids use modern technology combining thermal engines and electric motors—the latter supports the former during acceleration and when the vehicle requires greater driving torque. These cars cannot travel only on electric power; therefore, the combustion engine will always remain on throughout the journey.

Full hybrids are similar to mild hybrids but are equipped with a slightly more powerful electric motor. They can travel, if the battery charge level allows it, for a few kilometers in 100% electric mode.

The BEV is equipped with a lithium-ion battery pack. Inside the battery, the positively charged lithium ions are transported via a liquid electrolyte from the anode to the cathode via a separator. This movement causes the generation of an electric current. Unlike a car with an internal combustion engine, an electric car is not supplied with fuel, but with electricity which is stored in a battery pack. The capacity of the batteries determines the range of the vehicle.

In my opinion, to shorten charging times along the journey, it would be necessary to increase the energy capacity of the batteries or to equip service stations with charged spare batteries. Professor Mauro Marchionni[18] clearly illustrated the charging problems connected to the adaptation of the national electricity distribution network by operators.

Hydrogen fuel cell cars, on a construction level, do not need a large battery to store energy. While they move thanks to one or more electric motors, the difference from EVs lies in the presence of a hydrogen tank that powers a stack of fuel cells and a "buffer" battery. The role attributed to fuel cells and PEM membranes is therefore of primary importance. Electrical energy is obtained from a reaction between the oxygen and the hydrogen contained in the tank.

Already in 2001, as director of the Motors Institute, Giancarlo Michellone activated the testing room for hydrogen fuel cell electric powertrains, to conduct the project of the Seicento Elettra hydrogen fuel cell.

I went to Germany to test a fuel cell stack from Proton Motor Fuel Cell GmbH. Later that April, with Giancarlo Michellone, Director of the Fiat Research Center and member of the scientific council of the Motors Institute, he obtained substantial funding from the Ministry of the Environment. The project was aimed at the development of the Seicento Elettra H2 fuel cell with the collaboration of ENEA and the Universities of Turin, Rome, and Naples. The Fiat research center entrusted the resources to Fiat of Pomigliano d'Arco (then one of the hubs relating to telematics and control technologies). Fiat strongly advised the institutions to

---

[18] The Electric car, Mauro Marchionni, Claudio Monza.

prepare an infrastructure that could accommodate that type of car, reiterating that "the most critical aspect for the future fuel cell vehicle was represented by the hydrogen production, distribution, and storage infrastructure."

The Seicento Elettra H2 stack was powered by oxygen coming from a compressed air tank at 1.3 bar and by hydrogen coming from six small cylinders of nine liters each at a pressure of 1.5 bar. To make the system work, it was necessary to raise the voltage at the stack output from 48 to 216 volts for the operation of the 30-kilowatt traction motor.[19] For the refuelling, it took around 10 minutes to recharge the hydrogen cylinders. An important result of the project was the installation of a device called a reformer onboard future vehicles, which allows hydrogen to be obtained directly from methane. Figure 4.3 shows the schematic of an EV with a compressed green hydrogen fuel cell.

FIGURE 4.3 EV with compressed green hydrogen fuel cell.

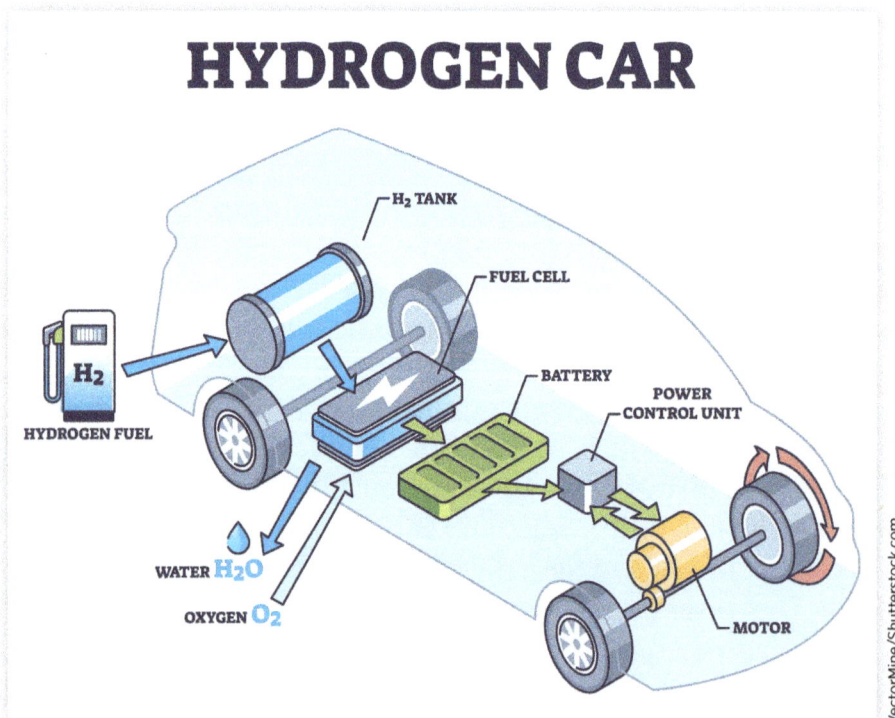

---

[19] P. Corbo, Felice E. Corcione and F. Migliardini. Conference organized by the director of the CNR Motors Institute, 2004.

Mega-Trends in Technology

Contribution from: Giuseppe Esposito Corcione
Founder and CEO of the Modena-based company "Reinova"

With fewer energy resources, more digital technology, and shifted individual habits, future investment plans should be channelled in an appropriate, efficient, and effective way.

This massive upcoming industrial revolution touches the conscience, the future, and the imagination of all of us. Technological progress should result in a constant improvement in the quality of life of every single living form. To progress technologically, we need to analyze cultural changes and modern human needs. The determining factor for new technological areas is time. We must give back time (and quality of time) to the people by supporting humanity with technology while maintaining respect for fundamental laws that take nature into balance.

The new technological challenges are based on three major pillars: profitability, adaptivity, and sustainability.

To progress we must innovate; to innovate we must have the ability and courage to imagine—to curiously seek new paths leading us to optimize every single link and detail of a chain that contributes to improving every single daily action.

Profitability means developing systems based on machine learning and artificial intelligence capable of connecting us, things, and infrastructure in real time; anticipating needs with the aim of convergence; optimizing time; and minimizing effort. Profitability should enhance the plurality that each individual can express, by helping to generate points of convergence. This applies to the industrial world, to mobility of the future, to the management of services, and to the control of production demand.

The environmental footprint of the electric car will be reduced as the entire energy production chain follows the ecological transition, reducing greenhouse gas and harmful emissions through renewable sources. Ultimately, "sustainable" means electric traction, whether its energy source is a battery, hydrogen fuel cell, or other form.

Being connected means being able to transfer data—lots of it—in real time. In the context of personal mobility, faster data transfer can allow for an optimized experience between a driver and other systems and vehicles. The constant search for connectivity opens new scenarios, such as computer security (e.g., mobility cybersecurity) and the development of human-machine interface systems that allow humans to interact intelligently with various machines in real time.

Shared means that the vehicle must maximize the use and accommodate the different needs of drivers, as well as responding to their on-demand requests. The cloud and predictive development will help to optimize this aspect by radically changing the traditional sense of ownership and by pushing car manufacturers to increasingly become service and software development companies. In the meantime, vehicles will become commodities.

Technological progress should be followed by training for new generations, so to arouse their curiosity and to lead them to always seek new solutions and ideas.

The technological challenges of tomorrow are also based on the ability to adapt learning and teaching to encourage the creation of a stimulating ecosystem for each individual student.

Technology and training are essential for technological progress, but we should not forget that it is essential to stimulate dreams and imagination, without which innovation could not exist.

Nature challenges us to be in solidarity and attentive to the protection of creation, also, to prevent, as far as possible, the most serious consequences.

—*Pope Francis*

Epilogue

The desire to write this book came from the great commitment and constant participation in addressing problems regarding the environment and sustainable mobility. Throughout my life I have always reconciled my scientific and personal activities with social commitment aimed at achieving an ecologically sustainable planet.

There have been many international, national, and local meetings to raise awareness about the problem of sustainable mobility.[1] Already in 1999, as previously reported, with the participation of the Fiat Research Center, my friends and colleagues from the Universities of Naples, Turin, and Milan, created the first fuel cell electric car, the Fiat Seicento Elettra H2, for the green hydrogen company Nuvera.

In 2003, I gave a speech at the international conference on the environment, waste, water, energy, and organic living. In August 2022, at the invitation of Dr. Prisco Piscitelli of the Italian Society of Environmental Medicine and Professor Annarita Corrado, head teacher of the Leonardo da Vinci scientific high school in Maglie, I participated in the seminar on the environment and green mobility in the presence of teachers, students, parents, and local and national television networks.

---

[1] Felice E. Corcione, "International Conference on the Environment: Waste, Water, Energy Living Bio," April 9–12, 2003, Catania.

Epilogue

It took many months to complete this book, because of the topics covered and the continuous variations in the proposals and actions of the European Union on sustainable mobility. I published this book on June 5, 2023, on the World Environment Day established at the meeting of the United Nations in Stockholm on December 15, 1972, with the aim of drastically reducing the use of plastics as already agreed at the G7 in Sapporo of April 2023.

In conclusion, on April 22, 2023, World Earth Day, my friends and fellow researchers and I meet once again in Prata di Principato Ultra (AV), in a fascinating setting in which the surrounding hills frame the bucolic landscape. The vines have not yet completely sprouted: the continuous rains, the unusual cold, and the hail have delayed their growth. April 22 seemed like the right day to me, as on this specific day the environmental protection of the planet went viral: the mass media gave great emphasis to the application of the "Three Rs" rule: Reduce, Recycle, and Reuse.

Our talk focuses on the main points discussed in the previous months: the causes of environmental degradation and the possible solutions using green hydrogen produced from renewable sources. Everyone agrees on the urgent need to decarbonize the planet by 2050. For green mobility, there is agreement on the use of fuel cell electric traction, biofuels, and synthetic e-fuels for internal combustion engines.

The solution for the future is of course nuclear fusion to produce hydrogen for civil and industrial uses and for zero-emission mobility.

Therefore, the energy transition should contemplate the best possible use of an energy mix that does not exclude fossil fuels, biofuels, and green hydrogen. However, many years of research, development, and financial investments are still necessary to make the Italian and global energy system stable and intelligent.

Our hashtag should be:
THINK GREEN TO SAVE THE PLANET.

Conferences and Articles by the Author and Colleagues

Appendix A: Catania 2003: Electric-Fuel Cell Vehicles for Sustainable Mobility

Felice E. Corcione

Director of Istituto Motori – CNR Napoli

Università di Catania 9–12 Aprile 2003
Conferenza Internazionale Sull'Ambiente

Abstract

Hydrogen, the energy carrier of the future, used in fuel cells, will allow us to have "zero emissions" vehicles (cars, buses, scooters, etc.). However, to make this technology a reality, the problems associated with the production and distribution of hydrogen should be solved. In this speech, the issues related to the evolution of propulsion systems and fuels for sustainable mobility will be discussed: the internal combustion engine, fuel, fuel cells.

Appendix B: Bressanone 2005: Exhibition and Test of Electric Vehicles

Felice E. Corcione, *Istituto Motori CNR Napoli*

The need for a solution to the problem of air and noise pollution in urban areas has led to the development of advanced propulsion systems, with high efficiency and reduced local emissions. In this regard, vehicles with electric propulsion offer a possible alternative to traditional propulsion with a combustion engine, in terms of performance, noise and efficiency, and pollution with this type of vehicle is zero in the areas of use. The main problems linked to the diffusion of electric vehicles are caused by the still reduced autonomy of the electrochemical accumulators on board and by the lack of availability of charging stations, which could find a first and simple installation in the main public parking areas. These critical issues seem to be resolvable through the use of fuel cells powered by hydrogen, which allow electricity to be produced directly on board. The advantages lie in the possibility to increase the autonomy of current battery-only electric vehicles, thus making the new hydrogen fuel cell vehicles comparable, in terms of performance, with traditional internal fuel vehicles of the same category. Furthermore, the use of hydrogen leads to high energy efficiency, with advantages in terms of energy saving, combined with the emission of only water vapor at the exhaust. For this reason, historic centers will be the first ones to benefit from the diffusion of technologies linked to the use of hydrogen.

So far, the attention of the large automotive vehicle brands has been directed to prototypes of fuel cell electric cars, but there is now the growing idea that two-wheeled vehicles equipped with hydrogen fuel cells will be the first hydrogen vehicles to be widely used in private. This is related to the greater simplicity of the vehicle and to the need to use small fuel cells, which are easier to manage and cost-effective. Unfortunately, even if few ones admit it, the problem of hydrogen technology is not in the engine technology, the improvement of which is proceeding rapidly, as it will also be demonstrated by various reports that will be presented at the Interactive Seminar in Bressanone, but precisely in the infrastructural availability of the fuel hydrogen, such as hydrogen tanks and refuelling stations.

As part of the 14th Interactive Seminar on Electric Drives - Technological Evolution and Emerging Issues, which was held in Bressanone at the Hotel Elefante on March 23 to 26, 2003, a presentation of two and four-wheeled electric road vehicles was organized. The presentation took place in two successive phases: one aimed at the citizens of Bressanone, who could personally try the vehicles with electric propulsion on display on Sunday March 23, 2003, the other one aimed mainly at the technical personnel of the sector, who took part of the seminar. The seminar, as usual, was aimed at an audience of Electric Drive experts operating in the research field and coming from various Italian and foreign universities and research bodies, as well as manufacturers with consolidated experience on the Italian and European markets.

Manufacturers in the sector were present at the demonstrations and showed great interest in the new technologies presented.

They were present:

Oxygen with a strategic base in Padua and an operational network in the North-East, leader in the production of electric scooters and pedal-assisted bicycles.

Always going against the trend, it invests most of its resources in research and development, as well as in the use of particularly innovative materials.

Oxygen has designed and created an electric scooter approved at European level in Italy.

The Oxygen scooter summarizes all the advantages of the zero-emission two-wheeled vehicle, combined with the need for very low management and maintenance costs.

FAAM S.p.A., manufacturer of traction and industrial batteries (ISO 14000 certified) since 1974, is a European leader in the sector of commercial electric vehicles approved for road circulation. The FAAM Group is present with three productions units in Italy. The flagship of FAAM production is the Camaleo Fuel Cell - powered by hydrogen - which uses PEM (Proton Exchange Membrane) technology, autonomy of around 90 kilometers with a charging cost of about €2. The application of the Fuel Cell project on the Camaleo electric bicycle is only the first step that will lead all FAAM electric vehicles to work with hydrogen fuel cells soon.

MICRO-VETT, born in 1987, is today the largest Italian manufacturer - and one of the first in Europe - of electric traction vehicles. Associated with ANFIA (National Association of Automobile Manufacturers) since 1992, it has developed the Electric Porter, by far the most widespread electric-powered four-wheeler in Italy with over 2200 vehicles marketed for about 350 customers. Today Micro-Vett is oriented toward the production of larger vehicles, such as the electric/hybrid Ducato developed with the Fiat Research Center, the pure electric Daily 35-65 q, the Micro-Vett 35 q on an Isuzu chassis, vehicles fitted with batteries latest generation nickel-sodium (also available on the Porter), as well as the small car segment with the Casalini Ydea Electric, also equipped with high-performance batteries, with an effective autonomy of more than 150 kilometers in an urban cycle.

Micro-Vett, in concert with the administration of the Municipality of Reggio Emilia and other local institutional actors, has developed a sustainable mobility project for the solution of environmental problems.

Appendix C: International Journal of Environmental Research and Public Health

Prisco Piscitelli, Barbara Valenzano, Emanuele Rizzo, Giuseppe Maggiotto, Matteo Rivezzi, Felice E. Corcione, and Alessandro Miani

The manufacturers present at the demonstration allowed the participants to try the two and four-wheeled electric transport vehicles, to verify their performance in terms of ease of driving, very low noise and efficiency.

Abstract

Background: The Italian Society of Environmental Medicine has performed a preliminary assessment of the health impact attributable to road freight traffic in Italy. Methods: We estimated fine particulate matter (PM10, PM2.5) and nitrogen oxides (NOx) generated by road transportation of goods in Italy considering the number of trucks, the emission factors and the average annual distance covered in the year 2016. Simulations on data concerning Years of Life Lost (YLL) attributable to PM2.5 (593,700) and nitrogen oxides NO_2 (200,700) provided by the European

Environmental Agency (EEA), were used as a proxy of healthcare burden. We set three different healthcare burden scenarios, passing from 1/5 to 1/10 of the proportion of the overall particulate matter attributable to road freight traffic in Italy (about 7% on a total of 2262 tons/year). Results: Road freight traffic in Italy produced about 189 tons of PM10, 147 tons of PM2.5 and 4125 tons of NOx in year 2016, resulting in annual healthcare costs varying from 400 million up to 1.2 billion EUR per year. Conclusion: Road freight traffic has a relevant impact on air pollution and healthcare costs, especially if considered over a 10-year period. Any solution able to significantly reduce the road transportation of goods, could decrease avoidable mortality caused by air pollution and related costs.

Keywords: road freight traffic; transportation of goods; air pollution; years of life lost; healthcare burden.

Appendix D: Maurisi Vineyard Biodynamic Method

The biodynamic method takes care of the processes nourishing the forces linked to the soil on one side and those similar to sunlight on the other side, with the goal of keeping them in harmony. Therefore, the vineyard "lives" in a balance between these two forces.

Biodynamics starts from the global knowledge of the planet and its relationship with the cosmos; this knowledge is not acquired overnight but only through a habit of observing nature and its laws. Today we are slowly acquiring a certain ecological knowledge, but we are still far from understanding life in all its manifestations.

The principles on which biodynamics is based were formulated by the Austrian Rudolf Steiner, the founder of anthroposophy, a conception of man and world, which—in the first quarter of last century—had brought a fertile renewal in the fields of medicine, pedagogy, of art and science in general, acquiring many followers throughout the Western world. Agriculture was the last sector treated by Steiner before his death and he did that on the request of farmers who saw (with concern) the first signs of degeneration and weakness characterizing the application of modern cultivation methods, particularly with the use of chemical fertilizer.

In Koberwitz, in 1924, he held eight lessons for farmers where the central theme was the health of the earth and the maintenance and increase of fertility to improve the quality of food. The course is full of practical advice. The most important one concerns the preparation of a maximum yield fertilizer. Composting and proper preparation are two fundamental points of this process. Steiner's insights regarding the fight against diseases and parasites will probably gain more and more importance as time goes on. The use of natural manure helps the formation of humus in the soil: a true reservoir of vitality and richness. Manure promotes processes implementing soil fertility. This is also associated with the use of pure water (i.e., rain and spring water), which becomes the living carrier of the "information" brought to the soil and the vines.

Appendix E: The Internal Combustion Engine

The internal combustion engine was officially created on May 13, 1854, when the English Office of the "Commissioners of Patents" issued a two-sheet printed document (order number 1072) in London, where Eugenio Barsanti, professor of physics and mathematics at the Ximeniano Institute, and Felice Matteucci, "Gentleman of Florence" in the Grand Duchy of Tuscany, described their patent and illustrated the operating principle of their machine:

> … it was preferred to use two cylinders with pistons and parts of the piston thus arranged in the individual cylinders. The required gaseous explosive compound is obtained by combining hydrogen gas with atmospheric air and the explosion of that compound is carried out by means of electricity intervening in the engine operation.

The combustion engine they designed (Figure E.1) is defined as it follows: a twin-cylinder, heavy atmospheric engine with deferred action on the return stroke of the piston, with counter-phase movement of the pistons. The fuel mixture is composed of hydrogen and air, and it is ignited by an electric spark, reproducing what was produced by the "Volta gun" experiment.

FIGURE E.1 Barsanti engine.

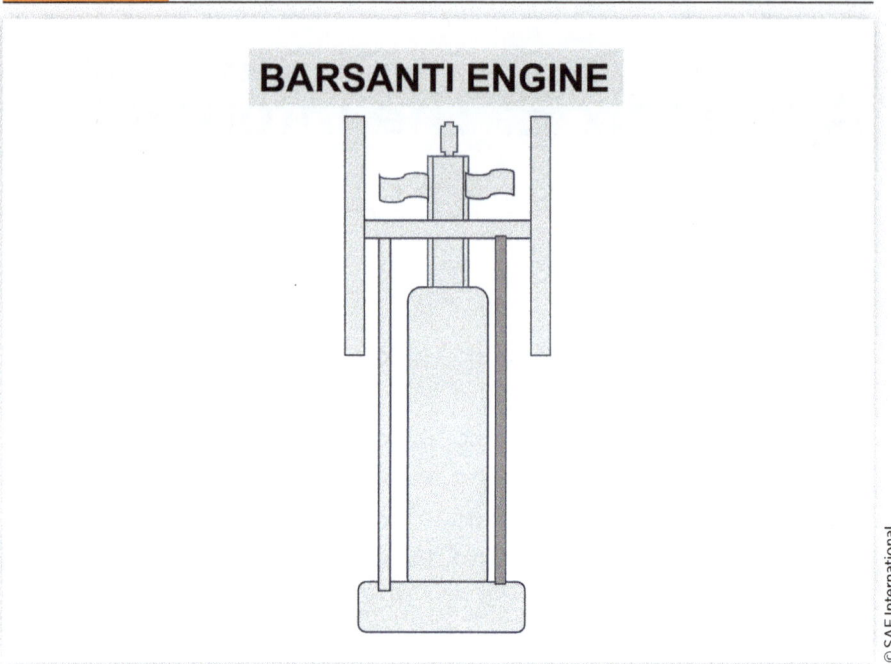

The operating cycle consists of intake, combustion, and exhaust, (i.e., a three-stroke engine). The driving thrust on the shaft is not caused by the expansion phase allowing the piston to rise (in the outward stroke), but by the descent phase through the action of atmospheric pressure and the weight force of the piston in the return stroke.

Nicolao Barsanti was born on October 12, 1821, in Pietrasanta, an ancient, glorious, and noble city in Versilia. At the age of 17, he changed his name to Eugenio and took on the religious habit of the Piarists. He belonged to a modest family. His father, Giovanni, like many of those who worked on the slopes of the great marble mountains, was a stonemason in a bust and statue workshop. His mother, Angela Francesconi, only carried out housework as it was customary during those times.

Eugenio carried out his first studies at the Institute of the Pious Schools of the Annunziata, more commonly known as The Congregation of the Scolopi, which was founded in 1617 by Giuseppe Calasanzio and elevated to a religious order in 1621. (It was suppressed in 1648 and then rehabilitated in 1669.)

The House of the Piarists in Pietrasanta was opened in 1819 during the reign of Ferdinand III of Lorraine with the Convent of Sant'Agostino. Barsanti, of frail constitution but of lively intelligence, achieved excellent results in his studies, revealing a particular predilection for mathematics and physics, so much so that—at only 20 years old and not yet ordained as a priest—he was sent to the Piarist College of San Michele in Volterra to teach.

He was especially passionate about the study of physics, and he personally worked to enrich the modest scientific cabinet of the school with instruments he made himself. In 1843, as he tells us in a manuscript document preserved at the Ximenian Observatory in Florence, while repeating the famous experience of the "Volta gun" before his students, he had the idea of applying it as a driving force—the first time that an explosion of a mixture of hydrogen and air, set on fire by an electrical device, might be used this way.

The apparatus used to carry out the "phlogopneumatic explosion" experiment which, as we read in some writings, had been given by Alessandro Volta, is kept in the Civic Museum of Volterra.

It is a copper ellipsoid equipped on one side with the usual electrical conductor, and on the other with a long neck intended to introduce the detonating mixture which then had to be closed with a cork. Barsanti observed that the apparatus produced greater heating the more the cap was pressed down, and that the values reached were maximum when the explosion occurred without being able to expel the cork itself. In this sense, the Tuscan physicist continued to study the phenomenon with Felice Matteucci. At the Ximenian Observatory, there is abundant preserved documentation, from which we can find notes and observations made relating to the experiments carried out through a gas cylinder equipped with a piston and a flywheel, where thundering mixtures of oxygen and hydrogen were ignited by the electric spark.

In 1849, he was called to Florence as a teacher of mathematics in the Casa of San Giovannino, where there was a school comparable to a modern scientific high school. Shortly afterwards, he was appointed Professor of Mathematics and Hydraulics in the Ximeniano Institute (now known as the Ximenian Observatory), which at that time was comparable to a university.

In Florence, Barsanti, encouraged by his brothers Giovanni Antonelli and Filippo Cecchi, continued his studies and experiences which led him to interesting scientific observations. Unfortunately, he lacked the applied mechanical skills necessary for the practical creation of a mechanical device that could use his intuitions about the internal combustion engine.

At the end of 1851, he established the collaboration with Felice Matteucci, a profound expert in mechanics and frequenter of the Ximenian Observatory, and the two developed a solid friendship. Barsanti communicated his ideas and studies to him, and Matteucci put them into practice thanks to his excellent mechanical knowledge.

The studies, continued jointly in Florence, tended

> ...to conceive a mechanism that would not only renew the introduction and ignition of the detonating mixture very frequently and in the right quantity, but that would also reject all the products and residues of combustion to the outside and, above all, that would release the pistons in their movement and connect them promptly and with stability...,

as it was stated in the professors' report of Giovanni Codazza, Camillo Hajech, and Luigi Magrini in the proceedings of the session of July 22, 1863, at the Lombard Institute of Sciences, Letters and Arts.

After tenacious and patient studies, the two inventors finally gave substance to their idea. At the conclusion of a long series of experiments and laborious research, they created the first engine.

Figure E.2 shows the engine, patented by Eugenio Barsanti and Felice Matteucci, as it is exhibited in the Ximenian Museum. The report also suggests the possible applications of this engine to the movement of ship blades and to any mechanical means requiring a driving force.

FIGURE E.2 Barsanti and Matteucci engine.

Index

A

Acerra Waste-to-Energy power station, 8
Adaptivity, 51
Agriculture, 9, 24, 67
Air pollution, 4, 24–25
Ammonia (NH$_3$), 36–38
Anaerobic digestion, 41

B

Barsanti and Matteucci engine, 69–73
Battery electric vehicles (BEVs), 10–11, 49
Biodynamic method, 25, 67–68
Biofuels, 24, 26–27, 45
Biogas, 26, 41
Biomethane, 26, 41
Bloom Energy, 46
Blue hydrogen, 35
Bressanone 2005, 61–63

C

Catania 2003, 59
Circular economy, 28–29
CO$_2$ emissions, 3, 17, 19, 35
Cummins Inc., 46

D

Diesel, 10–11, 26
Dual-fuel injection system, 47

E

"E turn," 2
Ecofining™ system, 26
Electrification of mobility, 43
Electronic fuel (E-fuel), 27
Environmental degradation
  agriculture, 9
  air pollution, 4
  causes of, 2, 5–6
  conditions, 1, 2
  diesel and electric cars, 10–11

economic models, 3
EEA, 2–3
green mobility, 3
landfills, 7–8
municipal solid waste, 6–7
transport, 9–10
waste-to-energy plants, 8–9
zero-emission mobility, 4
European Environment Agency (EEA), 2–3
European Green Deal, 28
European Union Proposals
  actions, 16–18
  Euro 7 Regulation, 18–19
  Fit for 55, 15–16
Euro 7 regulation, 18–19, 45

F

FAAM Group, 63
Fit for 55, 15–16
Friction braking systems, 5–6
Full hybrid models, 48

G

Greenhouse effect, 5–6
Green hydrogen, 34, 35
Grey hydrogen, 35

H

Honeywell UOP, 27
Human-machine interface systems, 52
Hydroelectric energy, 24
Hydrogen
  advantages, 44
  ammonia, 36–38
  anaerobic digestion, 41
  Bloom Energy, 46
  cost of solar, 39
  Cummins Inc., 46
  definition, 33

electrification of mobility, 43
energy revolution, 39
gasoline and natural gas, 34
IVECO Fiat Group, 44–45
low emission cars, 48–50
nuclear fusion, 40
PEM, 43, 44
storage and transport, 42
Toyota, 37
types of, 35
waste-to-energy plants, 40–41

I

Internal combustion engine, 69–73
International Journal of Environmental Research and Public Health, 65–66
International Renewable Energy Agency, 24
IVECO Fiat Group, 44–45

K

Kyoto Protocol, 3

L

Landfills, 7–8
Laser light scattering technique, 5
Leachate, 7–8
Light-emitting diodes (LEDs), 36
Linear economic model, 28
Low emission cars, 48–50

M

Maurisi Vineyard biodynamic method, 67–68
Mazda Motor Corporation, 42
Micro-Vett, 63
Mild hybrids, 48
Municipal solid waste, 6–7

© 2024 SAE International

N
NOx emissions, 18
Nuclear fusion, 40

O
Oxygen, 62–63

P
"Phlogopneumatic explosion" experiment, 71
Pink hydrogen, 35
Plug-in hybrids, 17, 48
Profitability, 51–52
Proper waste management, 6–7
Proton Exchange Membrane (PEM), 43, 44

R
Renewable energy
   A16 motorway, 23
   biofuels, 26–27
   circular economy, 28–29
   hydroelectric energy, 24
   photovoltaic, 23
   solutions and remedies, 24–25
   synthetic fuel and e-fuels, 27
Replanet Magazine, 29
Respiratory system, 6

S
Seicento Elettra H2 stack, 50
Soil pollution, 24–25
Solar energy, 39
Sustainability, 51
Synthetic fuel, 27

T
Technological progress, 51–52
Toyota, 37
Trans-esterification process, 26
Transportation, 9–10

V
Vehicle mobility, 5
"Volta gun" experiment, 69, 71

W
Waste electrical and electronic equipment (WEEE), 29
Waste-to-energy plants, 8–9, 40–41
Water pollution, 24–25

X
Ximenian Observatory, 71, 72

Y
Yellow hydrogen, 35

Z
Zero-emission mobility, 4, 17

About the Author

Biography of Felice Esposito Corcione

Courtesy of Felice E. Corcione.

Felice Esposito Corcione was born in Marigliano outside of Naples, Italy, into a humble peasant family as the last of five brothers and a sister. On July 28, 1973, he graduated with full marks in electrical engineering at the Federico II University of Naples. From April 1, 1974, to June 3, 1975, he did military service in the fourth "Bersagliere" armored infantry regiment in Legnano. (The Bersaglieri were not motorized, therefore they do not pollute.)

On June 5, 1975, he married his wife, Annamarial, and had two children: Carola in 1976 and Giuseppe in 1979.

In 1974, he was hired as a National Research Council researcher at the Motor Institute of Naples and began intense activity in the field of fluid dynamics and combustion of endothermic engines. He published numerous scientific articles at conferences and international journals in the transport sector. In 1997, he was appointed by the Ministry of University and Research to represent Italy in the International Energy Agency, as a member of the Executive Committee in the implementing agreement "Energy Conservation in Combustion" from 1997 to 2004. The same year, the president of the Society of Automotive Engineers (now SAE International) invited him to be part of the Board of Directors with the rank of SAE Fellow.

In December 2010, he founded the SAE Naples Section with the participation of Rodica Baranescu and Murli Iyer.

From 1995 to 2000 and 2005 to 2008, he served as an adjunct professor at the University of Wisconsin-Madison and also at Wayne State University in Detroit in 2010. In 2012, he was invited to present his research results at the University of Seoul in Korea. Corcione also served mayor Marigliano from 2005 to 2009.

In January 2012, he retired from the CNR and founded the "Vigna Maurisi" farm in Prata di Principato Ultra (AV) and started the production of PDO grapes "Greco di tufo and Aglianico" using the biodynamic method. In 2023, he closed the farm and retired due to cerebral ischemia.

Military Service

To confirm my commitment as an environmentalist, from April 1, 1974 to June 3, 1975 I did military service in the 4th Armored Infantry Regiment of the Legnano division of Milan. The Bersaglieri are not motorized, therefore they do not pollute. On March 31, 1974, I left the Central Station of Naples in the company of about thirty recruits at 9:30 pm, headed for Legnano via Milan. We arrived in Legnano at 9:30 on April 1st, where we were awaited by a military truck that immediately took us to the Cadorna barracks over the bridge over the Olona river. Once we arrived in the barracks square, I witnessed the recruits rushing to go to the only telephone booth to inform their family members that we had arrived at our destination. Thus, I began my adventure as a Bersagliere assigned to the command and services company of the battalion located on the left side of the barracks, right in front of the tank battalion.

About the Author

The "Legnano" Infantry division was a large unit of the Italian Army. Established in 1945 as the "bersaglieri Legnano" Infantry division, it was reconfigured as a mechanized brigade in 1975, with the name of the "Bersaglieri-Carristi Legnano" mechanized brigade, dependent on the III Army Corps of Milan and included in the "Centauro" armored division with the "Goito" and "Curtatone" brigades and the general command in Bergamo. The Unit was deployed with all its departments in Lombardy and had a strength of 4,760 men (272 officers, 630 non-commissioned officers and 3,858 enlisted men). The "Legnano" was involved in civil operations, such as the rescue of the earthquake victims in Friuli in 1976, in Irpinia in 1980 and in military interventions, such as the operation in Lebanon, Italcon in 1982, the 1991 Gulf War and in Somalia in 1993. It was dissolved in 1996 following the issuing of the new defense model of the Italian armed forces for the dissolution of the Warsaw Pact. However, it survived until December 31, 1997, as the "Legnano" support unit command with six dependent regiments, most obviously the reconsult, deprived of the transmission company as it was part of the 1st transmission regiment of Milan.

www.ingramcontent.com/pod-product-compliance
Lightning Source LLC
Chambersburg PA
CBHW040109100526
44584CB00029BA/3959